Bioprocessing of Biofuels

Bioprocessing of Biofuels

Prakash Kumar Sarangi and Sonil Nanda

CRC Press
Taylor & Francis Group
Boca Raton London New York

CRC Press is an imprint of the
Taylor & Francis Group, an **informa** business

First edition published 2021

by CRC Press
6000 Broken Sound Parkway NW, Suite 300, Boca Raton, FL 33487-2742

and by CRC Press
2 Park Square, Milton Park, Abingdon, Oxon, OX14 4RN

© 2021 Taylor & Francis Group, LLC

CRC Press is an imprint of Taylor & Francis Group, LLC

Library of Congress Cataloging-in-Publication Data

Names: Sarangi, Prakash Kumar, author. | Nanda, Sonil, author.
Title: Bioprocessing of biofuels / author by Prakash Kumar Sarangi and Sonil Nanda.
Description: First edition. | Boca Raton : CRC Press, 2021. |
Includes bibliographical references and index.
Identifiers: LCCN 2020037720 | ISBN 9780367340841 (hardback) |
ISBN 9780429323829 (ebook)
Subjects: LCSH: Biomass energy.
Classification: LCC TP339 .S366 2021 | DDC 662/.88--dc23
LC record available at https://lccn.loc.gov/2020037720

ISBN: 9780367340841 (hbk)
ISBN: 9780429323829 (ebk)

Typeset in Times LT Std
by KnowledgeWorks Global Ltd.

Contents

Preface

The major issues relating to environmental sustainability such as a heavy dependency on fossil fuels, increased greenhouse gas emissions, pollution, global warming and climate change have drawn much attention around the globe to seek alternative energy sources that have negligible environmental impacts and societal benefits. There is an immense interest in biofuels research around the world due to their massive potentials in addressing the above-mentioned environmental concerns. Biofuels have the potential to supplement the current and future energy demands by being blended with fossil fuels or replacing them completely as drop-in fuels in automobiles as well as heat and power industries. Waste biomass, primarily lignocellulosic biomass (e.g. agricultural crop residues, forestry biomass and energy crops) and microalgae, can act as some cheapest renewable biosources for the production of biofuels and biochemicals. The prime focus of this book is to shed light on the bioprocessing of biofuels, especially through microbial conversion technologies to recover and transform the inedible polysaccharides into hydrocarbons biofuels and bioenergy.

The major subject areas discussed in this book are bioprospecting of bioresources for biofuel production (Chapter 1) including specific topics on biomass pretreatment and saccharification (Chapter 2), bioethanol (Chapter 3), biobutanol (Chapter 4), biomethanol (Chapter 5), biohydrogen (Chapter 6), algal biofuels (Chapter 7) and microbial fuel cells (Chapter 8). This book provides introductory synopses on the above-mentioned topics for applications in several cross-disciplinary areas of biotechnology, (bio) catalysis, fermentation technology, bioprocess engineering, chemical engineering and environmental sciences with a common agenda on biofuels and bioenergy. We express our sincere thanks to Ms. Renu Upadhyay (Assistant Commissioning Editor) and Ms. Shikha Garg (Editorial Assistant) from CRC Press for their enthusiastic assistance while developing this book.

Dr. Prakash Kumar Sarangi
Scientist
Directorate of Research
Central Agricultural University
Imphal, Manipur, India
Email: sarangi77@yahoo.co.in
ORCID: https://orcid.org/0000-0003-2189-8828

Dr. Sonil Nanda
Research Associate
Department of Chemical and Biological Engineering
University of Saskatchewan
Saskatoon, Saskatchewan, Canada
Email: sonil.nanda@usask.ca
ORCID: https://orcid.org/0000-0001-6047-0846

About the Authors

Dr. Prakash Kumar Sarangi is a Scientist with specialization in Food Microbiology at the Central Agricultural University in Imphal, Manipur, India. He has a Ph.D. degree in Microbial Biotechnology from the Department of Botany at Ravenshaw University, Cuttack, India; M.Tech. degree in Applied Botany from the Indian Institute of Technology Kharagpur, India; and M.Sc. degree in Botany from Ravenshaw University, Cuttack, India. Dr. Sarangi's current research is focused on bioprocess engineering, renewable energy, biofuels, biochemicals, biomaterials, fermentation technology and post-harvest engineering and technology. He has more than 10 years of teaching and research experience in Biochemical Engineering, Microbial Biotechnology, Downstream Processing, Food Microbiology and Molecular Biology. He has served as a reviewer for many international journals and has authored more than 45 peer-reviewed research articles and 15 book chapters. Dr. Sarangi has edited the following books entitled "*Recent Advancements in Biofuels and Bioenergy Utilization*" (Springer Nature), "*Biorefinery of Alternative Resources: Targeting Green Fuels and Platform Chemicals*" (Springer Nature), "*Fuel Processing and Energy Utilization*" (CRC Press), and "*Biotechnology for Sustainable Energy and Products*" (I.K. International Publishing House Pvt. Ltd.). Dr. Sarangi serves as an academic editor in PLOS One journal. He is associated with many scientific societies as a Fellow Member (*Society for Applied Biotechnology*) and Life Member (*Biotech Research Society of India; Society for Biotechnologists of India; Association of Microbiologists of India; Orissa Botanical Society; Medicinal and Aromatic Plants Association of India; Indian Science Congress Association; Forum of Scientists, Engineers & Technologists; International Association of Academicians and Researchers; Hong Kong Chemical, Biological & Environmental Engineering Society; International Association of Engineers*; and *Science and Engineering Institute*.

Dr. Sonil Nanda is a Research Associate at the University of Saskatchewan in Saskatoon, Saskatchewan, Canada. He received his Ph.D. degree in Biology from York University, Canada; M.Sc. degree in Applied Microbiology from Vellore Institute of Technology (VIT University), India; and B.Sc. degree in Microbiology from Orissa University of Agriculture and Technology, India. Dr. Nanda's research areas are related to the production of advanced biofuels and biochemical through thermochemical and biochemical conversion technologies such as gasification, pyrolysis, carbonization, torrefaction, and fermentation. He has gained expertise in hydrothermal gasification of various organic wastes and biomass including agricultural and forestry residues, industrial effluents, municipal solid wastes, cattle manure, sewage sludge, food wastes, waste tires, and petroleum residues to produce hydrogen fuel. His parallel interests are also in the generation of hydrothermal flames for the treatment of hazardous wastes, agronomic applications of biochar, phytoremediation of heavy metal contaminated soils, as well as carbon capture and sequestration. Dr. Nanda has published over 95 peer-reviewed journal articles, 40 book chapters, and has presented at many international conferences. He is the editor of books entitled *"New Dimensions in Production and Utilization of Hydrogen"* (Elsevier), *"Recent Advancements in Biofuels and Bioenergy Utilization"* (Springer Nature), *"Biorefinery of Alternative Resources: Targeting Green Fuels and Platform Chemicals"* (Springer Nature), *"Fuel Processing and Energy Utilization"* (CRC Press), and *"Biotechnology for Sustainable Energy and Products"* (I.K. International Publishing House Pvt. Ltd.). Dr. Nanda serves as a Fellow Member of the *Society for Applied Biotechnology* in India, as well as a Life Member of the *Indian Institute of Chemical Engineers, Association of Microbiologists of India, Indian Science Congress Association*, and the *Biotech Research Society of India*. He is also an active member of several chemical engineering societies across North America such as the *American Institute of Chemical Engineers*, the *Chemical Institute of Canada*, the *Combustion Institute-Canadian Section*, and *Engineers Without Borders Canada*. Dr. Nanda is an Assistant Subject Editor of the *International Journal of Hydrogen Energy* (Elsevier) as well as an Associate Editor of *Environmental Chemistry Letters* (Springer Nature) and *Applied Nanoscience* (Springer Nature). He has also edited several special issues in renowned journals such as the *International Journal of Hydrogen Energy* (Elsevier), *Chemical Engineering Science* (Elsevier), *Biomass Conversion and Biorefinery* (Springer Nature), *Waste and Biomass Valorization* (Springer Nature), *Topics in Catalysis* (Springer Nature), *SN Applied Sciences* (Springer Nature), and *Chemical Engineering & Technology* (Wiley).

Bioprospecting and Bioresources for Next-Generation Biofuel Production

1

1.1 INTRODUCTION

The sustainable utilization of waste organic biomass is an attractive option for the production of carbon-neutral biofuels to mitigate greenhouse gas emissions and address the rising global energy demands (Nanda et al. 2016c). Biofuels from renewable and biogenic materials (e.g. agricultural crop residues, forestry biomass, algae, energy crops, municipal solid waste, food waste and cattle manure) provide a wide range of advantages such as mitigation of greenhouse gas emissions, supplementing energy security, waste valorization and reinvigorating rural economy (Nanda et al. 2015b; Okolie et al. 2020a). Lignocellulosic biomass comprises agricultural crop residues, forestry biomass and energy crops, which contain cellulose, hemicellulose and lignin as their key biopolymeric components (Azargohar et al. 2019; Okolie et al. 2020b).

Waste biomass can be converted to biofuels and biochemicals through biological conversion technologies (e.g. fermentation and anaerobic digestion) and thermochemical conversion technologies (e.g. pyrolysis, gasification, liquefaction, transesterification, torrefaction and carbonization) to process specific liquid, gaseous and solid biofuels (Azargohar et al. 2013; Nanda et al. 2016b; Parakh et al. 2020). A wide variety of biofuels such as bioethanol, biobutanol, biohydrogen, biomethane, syngas, bio-oil, biodiesel and biochar are gaining attention for research and development worldwide for their potential use as sustainable fuels for automobiles, industries as well as combined heat and power (Nanda et al. 2014b). This chapter aims to discuss different generations of biofuels along with their composition and properties. The production of biofuels utilizing lignocellulosic biomass and conversion processes are also described in this chapter.

1.2 DIFFERENT GENERATIONS OF BIOFUELS

Bioethanol production from different starch-based food crops and grains (e.g. potato, cassava, corn, wheat, etc.) via fermentation technologies is considered as the first-generation biofuel (Nanda et al. 2018). Bioethanol and biodiesel are two liquid biofuels, which can be blended with petrol and diesel, respectively, to reduce greenhouse gas emissions (Nayak et al. 2019; Nayak et al. 2020). The supply chain of food crops and grains as well as the global economy suffer dramatically because of the reduced food supply and rising food prices owing to their diversion to first-generation biofuel refineries (Nanda et al. 2015b). Moreover, the arable or cultivable land area is also found to be competitive in such a scenario of food *versus* fuel debate associated with first-generation biofuel production.

The second-generation biofuels, on the contrary, can be generated from lignocellulosic feedstocks, which have no direct competition with the human food supply chain or animal feed (Okolie et al. 2019). These materials are easy to procure, abundantly available and relatively cheap to sustain second-generation biofuel refineries. Moreover, second-generation biofuels produced from lignocellulosic biomass are more enviable as they are non-edible, renewable and pose no threat to food crops and arable lands. A few examples of lignocellulosic biomass are bagasse, stalk, peel, straw, shell, husk, stem, wood shavings, sawdust, etc., which originate as residues from agricultural harvesting and forests. Depending on the crop variety, geography, weather and climatic conditions, agricultural and harvesting practices operate round the year across the globe. Hence, enormous amounts of waste plant residues tend to be generating globally and annually. It has been reported that the global production of agricultural wastes reaches 1.4 billion tons annually (Saini et al. 2015).

The third-generation biofuels are produced from algal biomass (Yadav et al. 2019). Cultivable land is not required for the production of third-generation biofuel feedstocks (i.e. algae), which is a major advantage of third-generation biorefineries. Moreover, algae can grow on wastewater while leading to bioremediation, heavy metal removal, bioenergy production (i.e. in situ lipid accumulation), carbon sequestration and greenhouse gas mitigation (Ankit et al. 2020). Last but not least, fourth-generation biorefining is devoted to the applications of genetic engineering to alter biofuel feedstocks and microorganisms for techno-economically efficient biofuel production (Sarangi and Nanda 2019b). Although fourth-generation biorefineries aim to achieve higher biofuel production rates in short durations with less energy and cost input, secondary markets aim for byproducts while posing relatively low carbon footprints. Figure 1.1 illustrates all the four generations of biofuels.

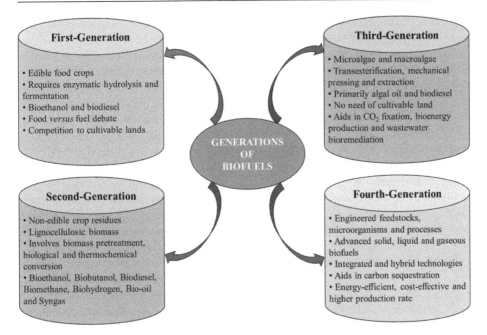

FIGURE 1.1   Different generations of biofuels

1.3 LIGNOCELLULOSIC BIOMASS AND PRETREATMENT

Lignocellulosic biomass is the most predominant organic matter on earth and has a wide range of industrial applications such as in the production of biofuels, biomaterials, biomaterials, nutraceuticals, pharmaceuticals, cosmeceuticals and carbon-based specialty materials (Okolie et al. 2020b). As mentioned earlier, lignocellulosic biomass constitutes agricultural crop residues, forestry residues, energy crops and invasive crops (Nanda et al. 2016a; Singh et al. 2020). Such biomasses are a ubiquitous source of renewable bio-based energy, which can favorably affect the sustainability matrix in terms of economics, employment, environmental concerns and energy security (Isikgor and Becer 2015). Considering their massive applications and utility, lignocellulosic biomasses are considered least deployed reservoirs of renewable natural polymers such as cellulose, hemicellulose and lignin. Lignocellulosic biomass typically contains 35–55 wt% cellulose, 20–40 wt% hemicellulose and 10–25 wt% lignin along with certain amounts of extractives and mineral matter (ash) (Nanda et al. 2013). Cellulose is composed of glucose (hexose sugar) monomers linked with β-1,4 glycosidic bonds, hydrogen bonds and Van der Waals force. On the other hand, hemicellulose comprises pentose sugars (e.g. xylose and arabinose) and hexose sugars (e.g. glucose, rhamnose, galactose and mannose) as well as sugar acids (e.g. glucuronic acid and galacturonic acid) (Nanda et al. 2015a). In contrast, lignin is a phenylpropane polymer having *p*-coumaryl alcohol, coniferyl alcohol and sinapyl alcohol as its basic building blocks (Fougere et al. 2016; Rana et al. 2018).

Due to the complex chemistry and interconnected network of cellulose, hemicellulose and lignin in lignocellulosic biomass, a pretreatment process is necessary before biological conversion. The pretreatment technologies involve physical agents (e.g. grinding, ozonolysis, ultrasound, microwave, etc.), chemical agents (e.g. acids, alkalis, organosolv, ionic liquids, liquid ammonia, steam, etc.) and biological agents (e.g. cellulolytic, lignin-degrading enzymes, lignin-modifying enzymes and microorganisms) (Nanda et al. 2014a; Nanda et al. 2014b).

Microorganisms consume organic substrates and utilize them in their metabolic processes, thereby generating useful products (metabolites), which can be further recovered as fuels, chemicals, nutraceuticals, pharmaceuticals, cosmeceuticals, flavoring agents, pigments, aromatic compounds and other value-added products with vast commercial applications (Sarangi and Nanda 2019a; Bhatia et al. 2020; Sarangi and Nanda 2020). Among the several microbial-assisted biofuels and biochemicals produced, bioethanol, biobutanol, biohydrogen, biomethane (biogas) and biomethanol are the most widely explored ones (Nanda et al. 2014a; Nanda et al. 2017a; Nanda et al. 2017b; Nanda et al. 2017c; Nanda et al. 2017d; Sarangi et al. 2018; Sarangi and Nanda 2018; Sarangi et al. 2020; Nanda et al. 2020a).

The growth of soil microbial heterotrophs is supported by their efficiencies to undergo natural enzymatic hydrolysis of lignocellulosic materials as plant debris, which is also an important process for terrestrial carbon cycling. Moreover, this process plays an important role in plant-microbial interactions to convert lignocellulosic materials to carbonaceous materials as well as CO_2 and CH_4. A variety of glycoside hydrolases are involved in enzymatic hydrolysis of complex lignocelluloses. The glycoside hydrolases family includes cellulases, hemicellulases, pectin-degrading enzymes and lignin-degrading enzymes.

Following biomass pretreatment, the recovered sugars are fermented using specific microorganisms to produce the desired alcohol-based biofuels and biochemicals. While both fungal and bacterial species are involved in ethanol fermentation, butanol production is achieved by *Clostridium*-aided acetone-butanol-ethanol (ABE) fermentation (Nanda et al. 2017b). Similarly, certain methanogenic bacteria (e.g. *Methanobacterium* sp., *Methanobrevibacter* sp., *Methanococcus* sp., *Methanoculleus* sp., *Methanofollis* sp., *Methanogenium* sp., *Methanomicrobium* sp., *Methanosarcina* sp., etc.) are involved in biomethane production through anaerobic digestion (Rana and Nanda 2019), whereas anaerobic bacteria such as *Clostridium* sp., *Enterobacter* sp. and *Bacillus* sp. are involved in dark fermentation to produce biohydrogen (Sarangi and Nanda 2020).

1.4 MICROBIAL BIOMASS IN BIOENERGY PRODUCTION

Basler et al. (2018) have reported on the efflux pump i.e. on native resistance-nodulation-cell division (RND) acting on short-chain alcohols. *Pseudomonas putida* has gained attention as a potential microorganism in biorefinery owing to diversified catabolism and elaborated stability to various lethal materials (Udaondo et al. 2012). Furthermore, due to the diversity in features such as compliant metabolism, suppleness to noxious substances and flexibility for metabolic engineering, *P. putida* is considered as the benign microorganism for the fourth-generation biofuels production. *P. putida* has also the ability for the production of *n*-butanol after expressing the biosynthetic pathway from *Clostridium acetobutylicum* (Nielsen et al. 2009). Moreover, the engineered strain of *P. putida* was employed in a biphasic liquid extraction system to aid in the formation and down streaming of toxic compounds in fermenters (Schmitz et al. 2015; Basler et al. 2018).

    P. putida is considered as a potential microorganism for industrial production of bioethanol because of the following factors (Dos Santos et al. 2004):

    i. Native elaborated defense to different stressors including several solvents.
    ii. Genetic stability.
    iii. Competence to develop vigorously on complex substrates.
    iv. Generally regarded as safe (GRAS).

Certain species of *Clostridium* and *Pseudomonas* are also employed to produce platform biochemicals. *Pseudomonas, Enterobacter* and *Bacillus* also have the potential to express transesterification activities leading to biodiesel production (Singh et al. 2008; Escobar-Niño et al. 2014). Implementing the genetic engineering tools, *P. putida* is confirmed as a powerful biocatalyst for the production of a wide range of value-added compounds such as non-ribosomal peptides, rhamnolipids, polyketides as well as aromatic and non-aromatic compounds (Loeschcke and Thies 2015).

    Considering the versatile activities of microorganisms, a consolidated bioprocessing system for the production of biofuels, biochemicals and other essential bioproducts can lead to a circular economy with maximum utility and marketability of desirable products and coproducts while ensuring sustainability (Sarangi and Nanda 2019b). For example, activated sludge can provide effective support and flourish the growth of microalgae for third-generation biofuel production due to the consortia of a few plant growth-promoting bacteria (PGPB), e.g., *Azospirillum* sp., *Pseudomonas* sp., *Bacillus* sp., *Rhodococcus* sp. and *Acinetobacter* sp. (Cea et al. 2015). This can be accomplished by two methods such as the cultivation of microalgae and bacteria in a single process and pretreatment of wastewater with bacteria for better growth of microalgae. Bacterial pretreatment of wastewater provides a suitable environment for the growth of algal biomass.

1.5 CONCLUSIONS

The conversion of waste biomass for sustainable fuel production is a major area of research and development around the world. Waste lignocellulosic biomass has the potential for biological and thermochemical conversion to produce biofuels, biochemicals and bioproducts. The first-generation biofuels have many limitations associated with food *versus* fuel and competition to cultivable lands. Nevertheless, strategies based on the utilization of ubiquitous lignocellulosic biomass, which is non-edible, seem to be more sustainable. A thriving lignocellulosic biorefinery can be comprehended by synergizing various technologies and biomass-processing approaches for sustainable and economical production of various clean fuels and platform chemicals. However, advancement in bioprocessing technologies and genetic engineering is a viable option to address many issues related to the commercial production of biofuels and other value-added products.

REFERENCES

Ankit, N. Bordoloi, J. Tiwari, S. Kumar, J. Korstad, K. Bauddh. 2020. Efficiency of algae for heavy metal removal, bioenergy production, and carbon sequestration. In: *Emerging Eco-Friendly Green Technologies for Wastewater Treatment*; ed. R.N. Bhargava; 77–101. Singapore: Springer Nature.

Azargohar, R., S. Nanda, A.K. Dalai, J.A. Kozinski. 2019. Physico-chemistry of biochars produced through steam gasification and hydro-thermal gasification of canola hull and canola meal pellets. *Biomass & Bioenergy* 120:458–470.

Azargohar, R., S. Nanda, B.V.S.K. Rao, A.K. Dalai. 2013. Slow pyrolysis of deoiled canola meal: product yields and characterization. *Energy & Fuels* 27:5268–5279.

Basler, G., M. Thompson, D. Tullman-Ercek, J. Keasling. 2018. A *Pseudomonas putida* efflux pump acts on short-chain alcohols. *Biotechnology for Biofuels* 11:136.

Bhatia, L., R.K. Bachheti, V.K. Garlapati, A.K. Chandel. 2020. Third-generation biorefineries: a sustainable platform for food, clean energy, and nutraceuticals production. *Biomass Conversion and Biorefinery*. Doi: 10.1007/s13399-020-00843-6.

Cea, M., N. Sangaletti-Gerhard, P. Acuna, I. Fuentes, M. Jorquera, K. Godoy, F. Osses, R. Navia. 2015. Screening transesterifiable lipid accumulating bacteria from sewage sludge for biodiesel production. *Biotechnology Reports* 8:116–123.

Dos Santos, V.A.P.M., S. Heim, E.R. Moore, M. Strätz, K.N. Timmis. 2004. Insights into the genomic basis of niche specificity of *Pseudomonas putida* KT2440. *Environmental Microbiology* 6:1264–1286.

Escobar-Niño, A., C. Luna, D. Luna, A.T. Marcos, D. Cánovas, E. Mellado. 2014. Selection and characterization of biofuel-producing environmental bacteria isolated from vegetable oil-rich wastes. *PLoS One* 9:e104063.

Fougere, D., S. Nanda, K. Clarke, J.A. Kozinski, K. Li. 2016. Effect of acidic pretreatment on the chemistry and distribution of lignin in aspen wood and wheat straw substrates. *Biomass & Bioenergy* 91:56–68.

Isikgor, F.H., C.R. Becer. 2015. Lignocellulosic biomass: a sustainable platform for the production of bio-based chemicals and polymers. *Polymer Chemistry* 6:4497–4559.

Loeschcke, A., S. Thies. 2015. *Pseudomonas putida*: a versatile host for the production of natural products. *Applied Microbiology and Biotechnology* 99:6197–6214.

Nanda, S., A.K. Dalai, J.A. Kozinski. 2014a. Butanol and ethanol production from lignocellulosic feedstock: biomass pretreatment and bioconversion. *Energy Science & Engineering* 2:138–148.

Nanda, S., A.K. Dalai, J.A. Kozinski. 2016a. Supercritical water gasification of timothy grass as an energy crop in the presence of alkali carbonate and hydroxide catalysts. *Biomass & Bioenergy* 95:378–387.

Nanda, S., A.K. Dalai, J.A. Kozinski. 2017a. Butanol from renewable biomass: highlights on downstream processing and recovery techniques. In: *Sustainable Utilization of Natural Resources*; eds. P. Mondal, A.K. Dalai; 187–211. Boca Raton, FL: CRC Press.

Nanda, S., D. Golemi-Kotra, J.C. McDermott, A.K. Dalai, I. Gökalp, J.A. Kozinski. 2017b. Fermentative production of butanol: perspectives on synthetic biology. *New Biotechnology* 37:210–221.

Nanda, S., J. Maley, J.A. Kozinski, A.K. Dalai. 2015a. Physico-chemical evolution in lignocellulosic feedstocks during hydrothermal pretreatment and delignification. *Journal of Biobased Materials and Bioenergy* 9:295–308.

Nanda, S., J. Mohammad, S.N. Reddy, J.A. Kozinski, A.K. Dalai. 2014b. Pathways of lignocellulosic biomass conversion to renewable fuels. *Biomass Conversion and Biorefinery* 4:157–191.

Nanda, S., J.A. Kozinski, A.K. Dalai. 2016b. Lignocellulosic biomass: a review of conversion technologies and fuel products. *Current Biochemical Engineering* 3:24–36.

Nanda, S., K. Li, N. Abatzoglou, A.K. Dalai, J.A. Kozinski. 2017c. Advancements and confinements in hydrogen production technologies. In: *Bioenergy Systems for the Future*; eds. F. Dalena, A. Basile, C. Rossi; 373–418. Cambridge: Woodhead Publishing, Elsevier.

Nanda, S., P. Mohanty, K.K. Pant, S. Naik, J.A. Kozinski, A.K. Dalai. 2013. Characterization of North American lignocellulosic biomass and biochars in terms of their candidacy for alternate renewable fuels. *Bioenergy Research* 6:663–677.

Nanda, S., R. Azargohar, A.K. Dalai, J.A. Kozinski. 2015b. An assessment on the sustainability of lignocellulosic biomass for biorefining. *Renewable and Sustainable Energy Reviews* 50:925–941.

Nanda, S., R. Rana, D.V.N. Vo, P.K. Sarangi, T.D. Nguyen, A.K. Dalai, J.A. Kozinski. 2020a. A spotlight on butanol and propanol as next-generation synthetic fuels. In: *Biorefinery of Alternative Resources: Targeting Green Fuels and Platform Chemicals*; eds. S. Nanda, D.V.N. Vo, P.K. Sarangi; 105–126. Singapore: Springer Nature.

Nanda, S., R. Rana, P.K. Sarangi, A.K. Dalai, J.A. Kozinski. 2018. A broad introduction to first, second and third generation biofuels. In: *Recent Advancements in Biofuels and Bioenergy Utilization*; eds. P.K. Sarangi, S. Nanda, P. Mohanty; 1–25. Singapore: Springer Nature.

Nanda, S., R. Rana, Y. Zheng, J.A. Kozinski, A.K. Dalai. 2017d. Insights on pathways for hydrogen generation from ethanol. *Sustainable Energy & Fuels* 1:1232–1245.

Nanda, S., S.N. Reddy, S.K. Mitra, J.A. Kozinski. 2016c. The progressive routes for carbon capture and sequestration. *Energy Science & Engineering* 4:99–122.

Nayak, S.K., B. Nayak, P.C. Mishra, M.M. Noor, S. Nanda. 2019. Effects of biodiesel blends and producer gas flow on overall performance of a turbocharged direct injection dual-fuel engine. *Energy Sources, Part A: Recovery, Utilization, and Environmental Effects*. Doi: 10.1080/15567036.2019.1694101

Nayak, S.K., P.C. Mishra, S. Nanda, B. Nayak, M.M. Noor. 2020. Opportunities for biodiesel compatibility as a modern combustion engine fuel. In: *Biorefinery of Alternative Resources: Targeting Green Fuels and Platform Chemicals*; eds. S. Nanda, D.V.N. Vo, P.K. Sarangi; 457–476. Singapore: Springer Nature.

Nielsen, D.R., E. Leonard, S.H. Yoon, H.C. Tseng, C. Yuan, K.L.J. Prather. 2009. Engineering alternative butanol production platforms in heterologous bacteria. *Metabolic Engineering* 11:262–273.

Okolie, J.A., R. Rana, S. Nanda, A.K. Dalai, J.A. Kozinski. 2019. Supercritical water gasification of biomass: a state-of-the-art review of process parameters, reaction mechanisms and catalysis. *Sustainable Energy & Fuels* 3:578–598.

Okolie, J.A., S. Nanda, A.K. Dalai, F. Berruti, J.A. Kozinski. 2020a. A review on subcritical and supercritical water gasification of biogenic, polymeric and petroleum wastes to hydrogen-rich synthesis gas. *Renewable and Sustainable Energy Reviews* 119:109546.

Okolie, J.A., S. Nanda, A.K. Dalai, J.A. Kozinski. 2020b. Chemistry and specialty industrial applications of lignocellulosic biomass. *Waste and Biomass Valorization*. Doi: 10.1007/s12649-020-01123-0

Parakh, P.D., S. Nanda, J.A. Kozinski. 2020. Eco-friendly transformation of waste biomass to biofuels. *Current Biochemical Engineering* 6:120–134.

Rana, R., S. Nanda. 2019. Application of methanation as an alternate energy technology. In: *Biogas Technology*; eds. S. Mishra, T.K. Adhya, S.K. Ojha; 105–116. New Delhi, India: New India Publishing Agency.

Rana, R., S. Nanda, V. Meda, A.K. Dalai, J.A. Kozinski. 2018. A review of lignin chemistry and its biorefining conversion technologies. *Journal of Biochemical Engineering and Bioprocess Technology* 1:2.

Saini, J.K., R. Saini, L. Tewari. 2015. Lignocellulosic agriculture wastes as biomass feedstocks for second-generation bioethanol production: concepts and recent developments. *Biotechnology* 5:337–353.

Sarangi, P.K., S. Nanda. 2018. Recent developments and challenges of acetone-butanol-ethanol fermentation. In: *Recent Advancements in Biofuels and Bioenergy Utilization*; eds. P.K. Sarangi, S. Nanda, P. Mohanty; 111–123. Singapore: Springer Nature.

Sarangi, P.K., S. Nanda. 2019a. Bioconversion of agro-wastes into phenolic flavour compounds. In: *Biotechnology for Sustainable Energy and Products*; eds. P.K. Sarangi, S. Nanda; 266–284. New Delhi, India: I.K. International Publishing House Pvt. Ltd.

Sarangi, P.K., S. Nanda. 2019b. Recent advances in consolidated bioprocessing for microbe-assisted biofuel production. In: *Fuel Processing and Energy Utilization*; eds. S. Nanda, P.K. Sarangi, D.V.N. Vo; 141–157. Boca Raton, FL: CRC Press.

Sarangi, P.K., S. Nanda. 2020. Biohydrogen production through dark fermentation. *Chemical Engineering & Technology* 43:601–612.

Sarangi, P.K., S. Nanda, D.V.N. Vo. 2020. Technological advancements in the production and application of biomethanol. In: *Biorefinery of Alternative Resources: Targeting Green Fuels and Platform Chemicals*; eds. S. Nanda, D.V.N. Vo, P.K. Sarangi; 127–139. Singapore: Springer Nature.

Sarangi, P.K., S. Nanda, R. Das. 2018. Bioconversion of pineapple residues for recovery of value-added compounds. *International Journal of Research in Science and Engineering* (CHEMCON Special Issue):210–217.

Schmitz, S., S. Nies, N. Wierckx, L.M. Blank, M.A. Rosenbaum. 2015. Engineering mediator-based electroactivity in the obligate aerobic bacterium *Pseudomonas putida* KT2440. *Frontiers in Microbiology* 6:284.

Singh, A., S. Nanda, F. Berruti. 2020. A review of thermochemical and biochemical conversion of *Miscanthus* to biofuels. In: *Biorefinery of Alternative Resources: Targeting Green Fuels and Platform Chemicals*; eds. S. Nanda, D.V.N. Vo, P.K. Sarangi; 195–220. Singapore: Springer Nature.

Singh, M., S. Singh, R.S. Singh, Y. Chisti, U.C. Banerjee. 2008. Transesterification of primary and secondary alcohols using *Pseudomonas aeruginosa* lipase. *Bioresource Technology* 99:2116–2120.

Udaondo, Z. 2012. Analysis of solvent tolerance in *Pseudomonas putida* DOTT1E based on its genome sequence and a collection of mutants. *FEBS Letters* 586:2932–2938.

Yadav, P., S.N. Reddy, S. Nanda. 2019. Cultivation and conversion of algae for wastewater treatment and biofuel production. In: *Fuel Processing and Energy Utilization*; eds. S. Nanda, P.K. Sarangi, D.V.N. Vo; 159–175. Boca Raton, FL: CRC Press.

Advanced Biomass Pretreatment Processes for Bioconversion

2

2.1 INTRODUCTION

On a global scale, the primary source of energy is heavily dependent on fossil fuels, especially crude oil, petroleum, diesel and natural gas (Rana et al. 2018a; Rana et al. 2019a; Rana et al. 2019b; Rana et al. 2020). Rapid urbanization, unprecedented industrialization and a continued upsurge in population growth are the key factors behind the rapid increase in the worldwide need for energy demand and depletion of natural energy sources (Nanda et al. 2016g). With the exploiting usage of fossil fuels as a primary source of energy and petrochemical derivatives in commercial products despite their rising prices, it is presumed that the world will face extreme challenges in the form of shortages of fossil fuels and experiential environmental concerns of greenhouse gas emissions, global warming and pollution (Singh et al. 2018; Shafiqah et al. 2020; Nguyen et al. 2020a; Nguyen et al. 2020b). In such scenarios, alternative renewable resources appear to reduce the adverse impact of greenhouse gas emissions and supplement the energy security and demands (Nanda et al. 2015b; Nanda et al. 2016e; Nanda et al. 2018a; Sarangi and Nanda 2020).

The potential renewable energy sources such as solar, wind, tidal, geothermal and biomass have been explored with multifarious applications. However, biomass-derived biofuels have the tendency to be used widely in the transportation sectors as well as industries and power plants for the combined heat and power generation (Azargohar et al. 2013; Nanda et al. 2014c; Azargohar et al. 2018; Azargohar et al. 2019; Kang et al. 2019; Kang et al. 2020; Parakh et al. 2020). Due to its abundant availability, cost-effectiveness and minimal impact on the food supply and cultivable lands, ligno-cellulosic biomass is a great candidate to produce biofuels to either partially or fully replace fossil fuel usage (Nanda et al. 2013; Nanda et al. 2017e; Okolie et al. 2020a; Okolie et al. 2020b). The exploration of sustainable fuel sources and essential chemicals from bioresources can be suitable options to mitigate the environmental problems associated with fossil fuels (Yadav et al. 2019). Fossil fuel is presumed to be substituted by biomass-based renewable energy by nearly 10–50% in 2030 (Cucchiella et al. 2014).

The priorities have been diverted towards enhancing biofuel production from renewable lignocellulosic biomass to address the issues of alternative fuels and greenhouse gas emissions. Although the first-generation feedstocks (e.g. corn, potatoes, cassava and other starch-based crops and grains) can be converted to biofuels, several factors restrict their utilization such as a threat to food security, competition to fertile lands and increased food prices (Muscat et al. 2020). Therefore, the focus has shifted towards utilizing non-food biomass like agricultural residues (Nanda et al. 2018b; Sun et al. 2020; Okolie et al. 2020c), forestry refuse (Nanda et al. 2016f; Nanda et al. 2017d), energy crops (Nanda et al. 2016c), invasive crops (Singh et al. 2020), animal manure (Nanda et al. 2016b), municipal solid wastes (Parakh et al. 2020), food waste (Nanda et al. 2016d; Nanda et al. 2019a), industrial effluents (Nanda et al. 2015c; Reddy et al. 2016), sewage sludge (Gong et al. 2017a; Gong et al. 2017b) and polymeric wastes (Nanda et al. 2019b) for producing next-generation biofuels, which have zero competition to the food chain. Among all the above-mentioned alternative resources, lignocellulosic feedstocks (i.e. non-edible plant biomass) has gained much attention for thermochemical conversion to bio-oil, biodiesel, synthesis gas and biochar (Mohanty et al. 2013; Nanda et al. 2016a) as well as biological conversion to bioethanol, biobutanol, biohydrogen and biomethane (Nanda et al. 2014a; Nanda et al. 2017b; Nanda et al. 2017a; Nanda et al. 2017c; Sarangi and Nanda 2018; Rana and Nanda 2019; Nanda et al. 2020). A wide variety of lignocellulosic feedstocks have been widely explored for alcohol-based biofuel and biochemical production after undergoing different pretreatment processes to facilitate microbial fermentation. The complexity of lignocellulosic biomass does not allow it to be directly fermented by microorganisms, thus demanding a suitable pretreatment. This chapter discusses the various strategies for biomass pretreatment processes.

2.2 BIOMASS PRETREATMENT TECHNOLOGIES

The natural recalcitrance of lignocellulosic biomass creates hurdles for microbes and/or their enzymes to convert cellulosic and/or hemicellulosic sugars to ethanol via fermentation (Sarangi and Nanda 2018; Sarangi and Nanda 2019b). This problem can be substantially solved by pretreating the biomass that not only accelerates the hydrolysis but also enhances the product yield (Sarangi and Nanda 2019a; Sarangi et al. 2020). The pretreatment process also removes certain physical and chemical barriers that prevent the access of enzymes and other hydrolytic agents to degrade the intricate cellulose-hemicellulose-lignin network. These barriers are responsible for rendering recalcitrance and inaccessibility towards enzymatic hydrolysis. A suitable pretreatment method removes hemicelluloses, depolymerizes lignin and hydrolyzes cellulose

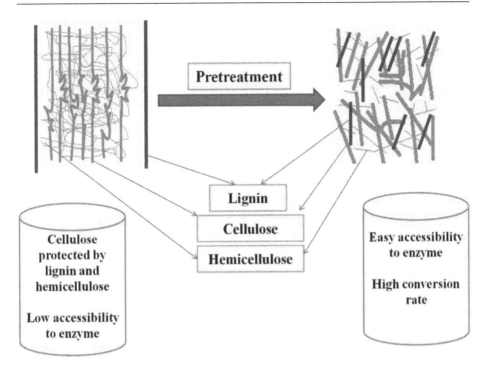

FIGURE 2.1 Schematic representation of biomass pretreatment

(Jönsson and Martín 2016). The schematic representation of a pretreatment process is shown in Figure 2.1. The accessibility of cellulolytic enzymes is enhanced after pretreatment, which is an outcome of alteration in pore size of biomass and reduction of cellulose crystallinity (Nanda et al. 2015a). The crystallinity index is a suitable method to authenticate the changes in the crystallinity of biomass by a pretreatment method.

Some typical biomass pretreatment strategies involve the use of mechanical forces (e.g. crushing and pulverizing), physical agents (e.g. ozonolysis, pulsed electrical field, gamma rays, electron beam, ultrasound and microwave), chemical agents (e.g. acids, alkalis, organosolv, ionic liquids, liquid ammonia, steam and subcritical water) and biological agents (e.g. cellulase enzymes, hemicellulase enzymes, lignin-modifying enzymes and Lignin-degrading enzymes) (Nanda et al. 2014b). However, the classification of pretreatments can also be reported as physicochemical, hydrothermal, chemimechanical, etc. depending on the integration of different pretreatment agents.

Several factors are responsible for the establishment of an ideal, cost-efficient and energy-efficient pretreatment process. Some key considerations and challenges encountered during biomass pretreatment are as follows (Nanda et al. 2014b; Maurya et al. 2015; Valdivia et al. 2016):

i. Enhancing the production of sugars (e.g. pentose and hexose) during enzymatic hydrolysis and saccharification from biomass.

ii. Curtailing further degradation of sugars and the formation of undesirable and inhibitory byproducts.

iii. Facilitating lignin recovery for its conversion into various value-added products.
iv. The flexibility of using reactors of moderate size to promote heat and mass transfer, recovery of products as well as minimizing the wastage of heat and power, thereby making the process cost-effective and energy-efficient.
v. Establishing an accomplished technology serving the purpose of achieving a significant volume of pretreated biomass irrespective of its type and source.
vi. Reducing the initial investments by adopting cost-effective reactor materials (e.g. steel alloys) that can tolerate corrosive acids and alkalis.
vii. Achieving cost-effective and eco-friendly methods that can perform best under ambient conditions.

2.3 PHYSICAL PRETREATMENT

Physical (also referred to as mechanical) pretreatment initially involves size reduction of biomass through chipping, milling and grinding to increase its surface area for effective action by other pretreating agents. Crushing and pulverizing the biomass ruptures cellulose fibrils and reduces its crystallinity (Fougere et al. 2016). Mechanical comminuting of biomass is a power-intensive process, which depends on the physical properties of biomass (i.e. initial size, moisture content, density and volume) as well as the desired final particle size.

Ultrasound is also identified as an effective approach to processing the waste biomass, which aids in saccharification processes along with the improvement in the extraction of hemicellulose, cellulose and lignin (Yang and Wang 2019). The need for cellulases can be curtailed after sonicating the biomass due to the depolymerization of its natural polymers (i.e. cellulose, hemicellulose and lignin). The reaction time involved in ultrasonic pretreatment is inversely proportional to the irradiation power applied (Imai et al. 2004).

2.4 CHEMICAL PRETREATMENT

A wide spectrum of acids, alkalis and oxidizing agents are involved in chemical pretreatment of lignocellulosic biomass. The effect of pretreatment on structural components varies on the type of chemicals employed. The removal of lignin becomes effective when alkalis, peroxides, oxides and ionic liquids are involved (Rana et al. 2018b). The recovery of hemicelluloses and cellulose becomes relatively easier after the depolymerization and removal of lignin from the biomass.

2.4.1 Dilute Acid Pretreatment

Dilute acid pretreatment is one of the widely used biomass pretreatment techniques. The solubilization of hemicellulosic, which is most effective because of dilute acid pretreatment, results in easy accessibility of cellulases for cellulose degradation (Antunes et al. 2019). High concentrations (e.g. 72% w/w H_2SO_4) and low concentrations (e.g. 0.4–4% w/w H_2SO_4) of acids are employed at elevated temperatures (120–200°C) for such a pretreatment process. This approach facilitates the efficient digestion of lignocellulosic biomass because of which less enzyme loading is needed during the subsequent saccharification step. However, in some instances, the enzyme-assisted hydrolysis stage is also avoided as fermentable sugars are generated effectively during acid pretreatment, which depends on the biomass properties including its particle size, acid concentration, temperature, reactor type and other process conditions (Zhu et al. 2009).

The concentration of acid (i.e. pretreating agent) also determines the generation of inhibitory substances like furfural, 5-hydroxymethylfurfural, phenols, acetic acid, carboxylic acids, etc., which tend to be inhibitory to the fermenting microorganisms (Nanda et al. 2014a; Sarangi and Nanda 2018; Sarangi and Nanda 2019b). For instance, the formation of furfural is reduced significantly to about three to five times with the use of dilute H_2SO_4 than with concentrated levels (Onoghwarite et al. 2016). The efficiency of mineral acids and organic acids like maleic acid and fumaric acid to pretreat lignocellulosic biomass has also been explored (Kootstra et al. 2009).

2.4.2 Alkaline Pretreatment

Alkaline pretreatment of biomass mostly involves alkaline hydroxides such as NaOH, KOH, $Ca(OH)_2$ and NH_4OH (Aswathy et al. 2010). The highly branched and cross-linked framework of lignin is damaged upon treatment with alkalis, thereby making it easier for cellulase and hemicellulases for biocatalysis (Rana et al. 2018b). The use of $Ca(OH)_2$ is a better approach over NaOH as the former is cost-effective, easily recoverable and relatively less corrosive (Mosier et al. 2005). Moreover, $Ca(OH)_2$ is also used for overliming the biomass hydrolyzates by neutralizing its acidity and removing the inhibitory degradation products such as phenols (Haq et al. 2018).

Alkaline pretreatment results in dramatic alterations to the lignin chemistry, distension of cellulose and solvation of hemicellulosic sugars (Sills and Gossett 2011). Cellulose is also de-crystallized to some extent during alkaline pretreatment (Goshadrou 2019). These effects are manifested due to the breakdown of ester and glycosidic chains along with the separations of structural connections between lignin and holocellulose (i.e. cellulose and hemicellulose). The internal surface area is also increased, which reduces the degree of polymerization of lignin, cellulose and hemicellulose as well as cellulose crystallinity because of alkaline pretreatment.

2.4.3 Wet Oxidation

Certain components in lignocellulosic biomass can be oxidized when reacted with an oxidant. The dissolution of hemicellulosic material along with the elimination of lignin content occurs because of wet oxidation. Typical compounds formed because of lignin decomposition during wet oxidation are CO_2, water and carboxylic acids.

Wet oxidation also aids in the removal of waxes and extractives. Bjerre et al. (1996) studied the wet oxidation of wheat straw with alkalis (20 g straw per liter at 170°C for 5–10 min), which resulted in 85% cellulose conversion to glucose.

2.4.4 Ionic Liquids

The exploration of ionic liquid seems to be a novel and potential technology for biomass pretreatment to generate fermentable sugars. Ionic liquids are organic salts with enhanced thermal stability having substantial application as green solvents in biomass degradation. Ionic liquids can potentially dissolve polar and non-polar organics, inorganics and polymeric compounds. The active dissolution of cellulose and hemicelluloses becomes feasible when biomass is pretreated with ionic liquids. Some advantages associated with the ionic liquids are solvent recycling, chemical stability, thermal stability (typically up to 400°C), non-flammability and non-volatility. Ionic liquids can dissolve solutes of fluctuating polarity. They are also involved in the production of novel chemicals and materials from biomass (Yoo et al. 2017).

Although ionic liquid-mediated biomass pretreatment is an appealing technology in recent years, its application at a large-scale needs to be evaluated in terms of the life cycle and techno-economic assessments. It is reported that the efficiency of ionic liquids gradually declines on their subsequent reuse (Liu et al. 2017). Another bottleneck associated with the use of ionic liquids is the requirement of higher temperatures (>100°C) and longer processing times for efficient biomass pretreatment. *Fusarium oxysporum* BN is an ionic liquid-tolerant fungus with the potential to produce ionic liquid stable cellulase enzyme. As reported by Xu et al. (2015), *F. oxysporum* BN can directly convert ionic liquid-pretreated rice straw to bioethanol with up to 0.125 g of ethanol per gram of rice straw.

2.5 PHYSICOCHEMICAL PRETREATMENT

2.5.1 Steam Explosion

Steam explosion is one of the traditional and widely used physicochemical pretreatments for biomass hydrolysis. Steam explosion is a mild, chemical-free (typically) and cost-effective approach to pretreat lignocellulosic biomass. Water is involved to heat the biomass under pressure followed by a sudden decompression of the reaction vessel to explode the cellulosic fibers (Pielhop et al. 2016). Temperature involved in this process varies from 160 to 260°C and pressure varies from 0.69 to 4.83 MPa (Yu et al. 2012).

The disruption of lignocellulosic biomass because of the steam explosion leaves behind the sugar polymers accessible to enzymatic saccharification. Hemicellulose degradation and transformation of lignin to pseudo-lignin are also the outcomes of the steam explosion (Vivekanand et al. 2013). Some factors such as residence time, temperature, biomass particle size and physicochemical properties can affect the efficiency of the steam explosion. As compared to other harsh pretreatment approaches, the steam explosion is a less reactive and less corrosive approach.

2.5.2 Liquid Hot Water

The pretreatment method involving liquid hot water requires water at higher temperatures (160–240°C) and pressures to retain its liquid state (Ruiz et al. 2013). This approach helps structural modification of lignocellulosic biomass along with the separation of hemicelluloses and amorphous cellulose, extraction of water-soluble components, although lignin remains unaffected (Nanda et al. 2018c). The quality of sugar generated is governed by the temperature employed while its quantity depends on the reaction time (Yu et al. 2010).

2.5.3 Ammonia Fiber Explosion

Ammonia fiber/freeze explosion (AFEX) method is a physicochemical process, which allows substantial delignification of biomass along with negligible degradation of sugars. AFEX is a cost-effective and less energy-intensive pretreatment process operating at moderate temperatures (60–90°C) and pressures (4–5 MPa) involving 1–2 kg of ammonia per kg of biomass (Peral 2016). Lower moisture content in the biomass and moderate conditions favor AFEX in an efficient pretreatment with lesser production of sugar degradation (or inhibitory) products. This process is somewhat similar to the steam explosion with the difference that liquid anhydrous ammonia is employed at elevated pressures and moderate temperatures followed by rapid de-pressurization (Chundawat et al. 2007).

2.5.4 Ammonia Recycle Percolation

In ammonia recycle percolation (ARP) strategy, nearly 5–15% of aqueous ammonia is permissible to percolate over a packed bed reactor retaining the biomass at a rate of about 5 mL/min (de Jong and Gosselink 2014). ARP is an improved approach over AFEX, due to its potential to remove 75–85% of lignin content and dissolve 50–60% of hemicellulose (Kim and Lee 2005). In this process, ammonia can be recycled. The pretreatment of corn stover and switchgrass with ARP resulted in the removal of 60–80% and 65–85% lignin, respectively (Iyer et al. 1996). Jonathan et al. (2017) researched dilute ammonia pretreatment of corn stover and reported the release of hemicelluloses from carbohydrate-lignin complex.

2.5.5 Hydrodynamic Cavitation

The approach of hydrodynamic cavitation accelerates the chemical reactions to pretreat a broad range of biomass with variable contents of cellulose, hemicellulose and lignin (Gogate 2016). The efficient removal of lignin along with the enhanced porosity of biomass is an outcome of this pretreatment method, which leads to effective saccharification of carbohydrates during enzyme hydrolysis. Other benefits associated with hydrodynamic cavitation are lower energy requirements and lower chemical catalyst loadings (Gogate and Pandit 2000).

The processing time of hydrodynamic cavitation is relatively shorter, which makes it an attractive option for a potential alternative to other pretreatments. Hydrodynamic cavitation has less complexity in its operating system, which can also be modified into a continuous process. Microbubbles are produced in the operating system when the fluid pressure declines and the fluid moves through a compression device such as a venture tube or an orifice plate (Patil et al. 2016). When microbubbles are generated, grown and collapsed, it gives rise to the cavitation phenomenon (Gogate and Pandit 2000). Inside the bubble, there exist drastic conditions such as high temperatures and pressures, which release a high amount of energy (Saharan et al. 2013). Along with these reactions, water molecules are detached in the cavities, thus ensuring the formation of potential oxidative radicals. The potential oxidative radicals are also liberated in the medium when the bubbles collapse (Badve et al. 2014). Due to the presence of these oxidative radicals, oxidation and degradation of lignin take place, thus resulting in its efficient elimination.

2.5.6 Supercritical Fluids

Supercritical fluids are a promising option for the pretreatment, fractionation, component extraction, structural modification, oxidation, liquefaction and gasification of waste organic biomass (Nanda et al. 2018c). Supercritical fluids behave likes gases and liquids, which open up their many applications. The critical temperature (T_c) and critical pressure (P_c) of water are 371°C and 22.1 MPa, respectively (Reddy et al. 2014). Water with its temperature and pressure lower or near to its critical points is called subcritical water, whereas that exceeding the critical points is called supercritical water. Subcritical and supercritical water are considered green solvents due to the innocuous nature of water and its abundance (Okolie et al. 2020a). Moreover, high-moisture containing biomass such as microalgae, sewage sludge and food waste can be effectively used in such pretreatments because no biomass drying is required since the reaction medium is water. This approach has the feasibility of application at large-scale industrial operations. Though this technology has been assessed in the pioneer stage, the last few years have witnessed it as a promising and green technology for pretreating, oxidizing, gasifying and incinerating different waste biomasses (Reddy et al. 2015; Reddy et al. 2017; Reddy et al. 2019).

Supercritical CO_2 is also used for biomass pretreatment, fractionation and biofuel upgrading (Reddy et al. 2018). The critical temperature and critical pressure of CO_2 is 31°C and 7.4 MPa, respectively, exceeding which it transforms into supercritical CO_2. The disruption of the crystalline structure of biomass occurs synchronously with an effective lignin elimination when the pretreatment process employs supercritical fluids. Subcritical CO_2 and subcritical water ameliorate the conversion of cellulose having the eco-friendly option as the usage of CO_2 curtails the impact of the greenhouse gas on the atmosphere (Liang et al. 2017). During this method, the moisture content of biomass plays a crucial role as CO_2 combines with water, thereby producing carbonic acid, which leads to hemicellulose hydrolysis.

2.6 BIOLOGICAL PRETREATMENT

Enzymatic hydrolysis involving a cocktail of enzymes such as β-glycosidases, cellulases, hemicellulases, lignin-modifying enzymes and lignin-degrading enzymes plays an important role in delignifying lignocellulosic biomass and depolymerizing cellulose and hemicellulose sugars for fermentation to fuels and chemicals (Álvarez et al. 2016; Kumar et al. 2016). Microorganisms producing cellulase differ from each other in possessing different components of cellulase, as there are multiple enzyme components present in a cellulose enzyme system. Based on the biocatalytic activity, cellulases are classified into endocellulases, exocellulases, cellobiases, oxidative cellulases and cellulose phosphorylases. The reducing and non-reducing ends are generated when endoglucanase works upon linear cellulose molecules. These ends are targeted by exoglucanase, which highlights more inner sites for endoglucanase binding. The products generated from the activity of exocellulases are further catalyzed by cellobiases or β-glycosidases. Radical reactions are catalyzed by oxidative cellulases, whereas phosphates are used to depolymerize cellulose in the biocatalysis aided by cellulose phosphorylase. *Pseudomonas aeruginosa* is an important microbial source of exoglycosidases (Kumar and Kumar 2017).

Many microorganisms have the potential to degrade cellulose, hemicellulose and lignin into their constitutive sugars. Prominent among them are bacteria, actinomycetes and fungi. These methods are a cost-effective and eco-friendly approach to obtaining the goals of pretreatment, thereby circumventing various challenges faced by other physical and chemical pretreatment methods. Through biological pretreatment, several constituents of biomass like lignin, cellulose, hemicellulose and polyphenols can be hydrolyzed or degraded (Sindhu et al. 2016). Glucose, arabinose, xylose, etc. are the mono sugars that are generated because of bacterial and fungal hydrolysis of cellulose and hemicellulose.

Selective degradation of lignin and hemicellulose occurs by the action of brown, white and soft-rot fungi, among which white-rot fungi are highly efficient. Biological pretreatments also occur in ambient temperatures and pressures, which significantly differs from other pretreatment methods in being environmental friendly. Furthermore, biological pretreatments do not require acid, alkali or other reactive and corrosive pretreating agents. Biological pretreatment methods are advantageous in being cost-effective and less energy-intensive, although they can be time-intensive. This limitation is associated with biological pretreatments because of the low rate of hydrolysis dependent on bacterial or fungal metabolism and enzymatic saccharification. Many researchers have explored a wide variety of microorganisms to explore lignocellulosic enzymes.

2.7 CONCLUSIONS

Bioconversion technologies towards efficient alcohol-based biofuel production involve novel pretreatment methods. Cost-effective, energy-efficient and sustainable pretreatment technologies are required for the industrial-scale conversion of waste biomass to recover fermentable sugars. The emergence of the new process technologies can lead to seeking sustainable alternatives for high-efficiency pretreatment of a wide variety of biomass. Novel and efficient pretreatment methods can be exploited for the production of biofuel that could meet the mounting energy demands. To get high-value biofuels and platform chemicals, more research on pretreatment technologies is required that can efficiently delignify lignocellulosic biomass, degrade holocellulose and prevent the formation of inhibitory and undesired degradation byproducts. Moreover, such pretreatment technologies should also have the flexibility for continuous-scale operations and feature short reaction time, less energy input, capital cost, less maintenance, safe operation, recycling of waste products and ability to be integrated into the bioconversion processes.

REFERENCES

Álvarez, C., F.M. Reyes-Sosa, B. Díez. 2016. Enzymatic hydrolysis of biomass from wood. *Microbial Biotechnology* 9:149–156.

Antunes, F.A.F., A.K. Chandel, R. Terán Hilares, A.P. Ingle, M. Rai, T.S. Milessi, S.S. da Silva, J.C. dos Santos. 2019. Overcoming challenges in lignocellulosic biomass pretreatment for second generation (2G) sugar production: emerging role of nano, biotechnological and promising approaches. *3 Biotech* 9:230.

Aswathy, U.S., R.K. Sukumaran, D. Lalitha, K.P. Rajeshree, R.R. Singhania, A. Pandey. 2010. Bio-ethanol from water hyacinth biomass: an evaluation of enzymatic saccharification strategy. *Bioresource Technology* 101:925–930.

Azargohar, R., S. Nanda, A.K. Dalai. 2018. Densification of agricultural wastes and forest residues: a review on influential parameters and treatments. In: *Recent Advancements in Biofuels and Bioenergy Utilization*; eds. P.K. Sarangi, S. Nanda, P. Mohanty; 27–51. Singapore: Springer Nature.

Azargohar, R., S. Nanda, B.V.S.K. Rao, A.K. Dalai. 2013. Slow pyrolysis of deoiled canola meal: product yields and characterization. *Energy & Fuels* 27:5268–5279.

Azargohar, R., S. Nanda, K. Kang, T. Bond, C. Karunakaran, A.K. Dalai, J.A. Kozinski. 2019. Effects of bio-additives on the physicochemical properties and mechanical behavior of canola hull fuel pellets. *Renewable Energy* 132:296–307.

Badve, M.P., P.R. Gogate, A.B. Pandit, C. Levente. 2014. Hydrodynamic cavitation as a novel approach for delignification of wheat straw for paper manufacturing. *Ultrasonics Sonochemistry* 21:162–168.

Bjerre, A.B., A.B. Olesen, T. Fernqvist. 1996. Pretreatment of wheat straw using combined wet oxidation and alkaline hydrolysis resulting in convertible cellulose and hemicellulose. *Biotechnology and Bioengineering* 49:568–577.

Chundawat, S.P.S., B. Venkatesh, B.E. Dale. 2007. Effect of particle size-based separation of milled corn stover on AFEX pretreatment and enzymatic digestibility. *Biotechnology and Bioengineering* 96:219–231.

Cucchiella, F., I. D'Adamo, M. Gastaldi. 2014. Financial analysis for investment and policy decisions in the renewable energy sector. *Clean Technology Environment Policy* 17:887–904.

de Jong, E., R.J.A. Gosselink. 2014. Lignocellulose-based chemical products. In: *Bioenergy Research: Advances and Applications*; ed. V.K. Gupta, M.G. Tuohy, C.P. Kubicek, J. Saddler, F. Xu; 277–313. Elsevier.

Fougere, D., S. Nanda, K. Clarke, J.A. Kozinski, K. Li. 2016. Effect of acidic pretreatment on the chemistry and distribution of lignin in aspen wood and wheat straw substrates. *Biomass & Bioenergy* 91:56–68.

Gogate, P.R. 2016. Greener processing routes for reactions and separations based on use of ultrasound and hydrodynamic cavitation. In: *Alternative Energy Sources for Green Chemistry*; ed. G. Stefanidis, A. Stankiewicz; London: Royal Society of Chemistry.

Gogate, P.R., A.B. Pandit. 2000. Engineering design methods for cavitation reactors II: hydrodynamic cavitation. *AIChE Journal* 46:1641–1649.

Gong, M., S. Nanda, H.N. Hunter, W. Zhu, A.K. Dalai, J.A. Kozinski. 2017a. Lewis acid catalyzed gasification of humic acid in supercritical water. *Catalysis Today* 291:13–23.

Gong, M., S. Nanda, M.J. Romero, W. Zhu, J.A. Kozinski. 2017b. Subcritical and supercritical water gasification of humic acid as a model compound of humic substances in sewage sludge. *The Journal of Supercritical Fluids* 119:130–138.

Goshadrou, A. 2019. Bioethanol production from cogongrass by sequential recycling of black liquor and wastewater in a mild-alkali pretreatment. *Fuel* 258:116141.

Haq, I., Y. Arshad, A. Nawaz, M. Aftab, A. Rehman, H. Mukhtar, Z. Mansoor, Q. Syed. 2018. Removal of phenolic compounds through overliming for enhanced saccharification of wheat straw. *Journal of Chemical Technology and Biotechnology* 93:3011–3017.

Imai, M., K. Ikari, I. Suzuki. 2004. High-performance hydrolysis of cellulose using mixed cellulase species and ultrasonication pretreatment. *Biochemical Engineering Journal* 17:79–83.

Iyer, P.V., Z.W. Wu, S.B. Kim, Y.Y. Lee. 1996. Ammonia recycled percolation process for pretreatment of herbaceous biomass. *Applied Biochemistry and Biotechnology* 57–58:121–132.

Jonathan, M.C., J. DeMartini, S. Van Stigt Thans, R. Hommes, M.A. Kabel. 2017. Characterization of non-degraded oligosaccharides in enzymatically hydrolysed and fermented, dilute ammonia-pretreated corn stover for ethanol production. *Biotechnology for Biofuels* 10:112.

Jönsson, L.J., C. Martín. 2016. Pretreatment of lignocellulose: formation of inhibitory by-products and strategies for minimizing their effects. *Bioresource Technology* 199:103–112.

Kang, K., S. Nanda, G. Sun, L. Qiu, Y. Gu, T. Zhang, M. Zhu, R. Sun. 2019. Microwave-assisted hydrothermal carbonization of corn stalk for solid biofuel production: optimization of process parameters and characterization of hydrochar. *Energy* 186:115795.

Kang, K., S. Nanda, S.S. Lam, T. Zhang, L. Huo, L. Zhao. 2020. Enhanced fuel characteristics and physical chemistry of microwave hydrochar for sustainable fuel pellet production via co-densification. *Environmental Research* 186:109480.

Kim, T.H., Y.Y. Lee. 2005. Pretreatment and fractionation of corn stover by ammonia recycle percolation process. *Bioresource Technology* 96:2007–2013.

Kootstra, A.M.J., H.H. Beeftink, E.L. Scott, J.P.M. Sanders. 2009. Comparison of dilute mineral and organic acid pretreatment for enzymatic hydrolysis of wheat straw. *Biochemical Engineering Journal* 46:126–131.

Kumar, A.K., B.S. Parikh, P. Mohanty. 2016. Natural deep eutectic solvent mediated pretreatment of rice straw: bioanalytical characterization of lignin extract and enzymatic hydrolysis of pretreated biomass residue. *Environmental Science and Pollution Research* 23: 9265–9275.

Kumar, R., P. Kumar. 2017. Future microbial applications for bioenergy production: a perspective. *Frontiers in Microbiology* 8:450.

Liang, J., X. Chen, L. Wang, X. Wei, H. Wang, S. Lu, Y. Li. 2017. Subcritical carbon dioxide-water hydrolysis of sugarcane bagasse pith for reducing sugars production. *Bioresource Technology* 228:147–155.

Liu, Z., L. Longfei, L. Cheng, X. Airong. 2017. Saccharification of cellulose in the ionic liquids and glucose recovery. *Renewable Energy* 106:99–102.

Maurya, D.P., A. Singla, S. Negi. 2015. An overview of key pretreatment processes for biological conversion of lignocellulosic biomass to bioethanol. *3 Biotech* 5:597–609.

Mohanty, P., S. Nanda, K.K. Pant, S. Naik, J.A. Kozinski, A.K. Dalai. 2013. Evaluation of the physiochemical development of biochars obtained from pyrolysis of wheat straw, timothy grass and pinewood: effects of heating rate. *Journal of Analytical and Applied Pyrolysis* 104:485–493.

Mosier, N., C.E. Wyman, B.D. Dale, R.T. Elander, Y.Y. Lee, M. Holtzapple, C.M. Ladisch. 2005. Features of promising technologies for pretreatment of lignocellulosic biomass. *Bioresource Technology* 96:673–686.

Muscat, A., E.M. de Olde, I.J.M. de Boer, R. Ripoll-Bosch. 2020. The battle for biomass: a systematic review of food-feed-fuel competition. *Global Food Security* 25:100330.

Nanda, S., A.K. Dalai, F. Berruti, J.A. Kozinski. 2016a. Biochar as an exceptional bioresource for energy, agronomy, carbon sequestration, activated carbon and specialty materials. *Waste and Biomass Valorization* 7:201–235.

Nanda, S., A.K. Dalai, I. Gökalp, J.A. Kozinski. 2016b. Valorization of horse manure through catalytic supercritical water gasification. *Waste Management* 52:147–158.

Nanda, S., A.K. Dalai, J.A. Kozinski. 2014a. Butanol and ethanol production from lignocellulosic feedstock: biomass pretreatment and bioconversion. *Energy Science & Engineering* 2:138–148.

Nanda, S., A.K. Dalai, J.A. Kozinski. 2016c. Supercritical water gasification of timothy grass as an energy crop in the presence of alkali carbonate and hydroxide catalysts. *Biomass & Bioenergy* 95:378–387.

Nanda, S., A.K. Dalai, J.A. Kozinski. 2017a. Butanol from renewable biomass: highlights on downstream processing and recovery techniques. In: *Sustainable Utilization of Natural Resources*; eds. P. Mondal, A.K. Dalai; 187–211. Boca Raton, FL: CRC Press.

Nanda, S., D. Golemi-Kotra, J.C. McDermott, A.K. Dalai, I. Gökalp, J.A. Kozinski. 2017b. Fermentative production of butanol: perspectives on synthetic biology. *New Biotechnology* 37:210–221.

Nanda, S., J. Isen, A.K. Dalai, J.A. Kozinski. 2016d. Gasification of fruit wastes and agro-food residues in supercritical water. *Energy Conversion and Management* 110:296–306.

Nanda, S., J. Maley, J.A. Kozinski, A.K. Dalai. 2015a. Physico-chemical evolution in lignocellulosic feedstocks during hydrothermal pretreatment and delignification. *Journal of Biobased Materials and Bioenergy* 9:295–308.

Nanda, S., J. Mohammad, S.N. Reddy, J.A. Kozinski, A.K. Dalai. 2014b. Pathways of lignocellulosic biomass conversion to renewable fuels. *Biomass Conversion and Biorefinery* 4:157–191.

Nanda, S., J.A. Kozinski, A.K. Dalai. 2016e. Lignocellulosic biomass: a review of conversion technologies and fuel products. *Current Biochemical Engineering* 3:24–36.

Nanda, S., K. Li, N. Abatzoglou, A.K. Dalai, J.A. Kozinski. 2017c. Advancements and confinements in hydrogen production technologies. In: *Bioenergy Systems for the Future*; eds. F. Dalena, A. Basile, C. Rossi; 373–418. Cambridge: Woodhead Publishing, Elsevier.

Nanda, S., M. Casalino, M.S. Loungia, A.K. Dalai, I. Gökalp, J.A. Kozinski. 2016f. Catalytic gasification of pinewood in hydrothermal conditions for hydrogen production. *Chemical Engineering Transactions* 50:31–36.

Nanda, S., M. Gong, H.N. Hunter, A.K. Dalai, I. Gökalp, J.A. Kozinski. 2017d. An assessment of pinecone gasification in subcritical, near-critical and supercritical water. *Fuel Processing Technology* 168:84–96.

Nanda, S., P. Mohanty, K.K. Pant, S. Naik, J.A. Kozinski, A.K. Dalai. 2013. Characterization of North American lignocellulosic biomass and biochars in terms of their candidacy for alternate renewable fuels. *Bioenergy Research* 6:663–677.

Nanda, S., R. Azargohar, A.K. Dalai, J.A. Kozinski. 2015b. An assessment on the sustainability of lignocellulosic biomass for biorefining. *Renewable and Sustainable Energy Reviews* 50:925–941.

Nanda, S., R. Azargohar, J.A. Kozinski, A.K. Dalai. 2014c. Characteristic studies on the pyrolysis products from hydrolyzed Canadian lignocellulosic feedstocks. *Bioenergy Research* 7:174–191.

Nanda, S., R. Rana, D.V.N. Vo, P.K. Sarangi, T.D. Nguyen, A.K. Dalai, J.A. Kozinski. 2020. A spotlight on butanol and propanol as next-generation synthetic fuels. In: *Biorefinery of Alternative Resources: Targeting Green Fuels and Platform Chemicals*; eds. S. Nanda, D.V.N. Vo, P.K. Sarangi; 105–126. Singapore: Springer Nature.

Nanda, S., R. Rana, H.N. Hunter, Z. Fang, A.K. Dalai, J.A. Kozinski. 2019a. Hydrothermal catalytic processing of waste cooking oil for hydrogen-rich syngas production. *Chemical Engineering Science* 195:935–945.

Nanda, S., R. Rana, P.K. Sarangi, A.K. Dalai, J.A. Kozinski. 2018a. A broad introduction to first, second and third generation biofuels. In: *Recent Advancements in Biofuels and Bioenergy Utilization*; eds. P.K. Sarangi, S. Nanda, P. Mohanty; 1–25. Singapore: Springer Nature.

Nanda, S., R. Rana, Y. Zheng, J.A. Kozinski, A.K. Dalai. 2017e. Insights on pathways for hydrogen generation from ethanol. *Sustainable Energy & Fuels* 1:1232–1245.

Nanda, S., S.N. Reddy, D.V.N. Vo, B.N. Sahoo, J.A. Kozinski. 2018b. Catalytic gasification of wheat straw in hot compressed (subcritical and supercritical) water for hydrogen production. *Energy Science & Engineering* 6:448–459.

Nanda, S., S.N. Reddy, H.N. Hunter, D.V.N. Vo, J.A. Kozinski, I. Gökalp. 2019b. Catalytic subcritical and supercritical water gasification as a resource recovery approach from waste tires for hydrogen-rich syngas production. *The Journal of Supercritical Fluids* 154:104627.

Nanda, S., S.N. Reddy, H.N. Hunter, I.S. Butler, J.A. Kozinski. 2015c. Supercritical water gasification of lactose as a model compound for valorization of dairy industry effluents. *Industrial & Engineering Chemistry Research* 54:9296–9306.

Nanda, S., S.N. Reddy, S.K. Mitra, J.A. Kozinski. 2016g. The progressive routes for carbon capture and sequestration. *Energy Science & Engineering* 4:99–122.

Nanda, S., S.N. Reddy, Z. Fang, A.K. Dalai, J.A. Kozinski. 2018c. Hydrothermal events occurring during gasification in supercritical water. In: *Supercritical and Other High-Pressure Solvent Systems: For Extraction, Reaction and Material Processing*; eds. A.J. Hunt, T.M. Attard; 560–587. London, Royal Society of Chemistry.

Nguyen, D.T., V.H. Nguyen, S. Nanda, D.V.N. Vo, V.H. Nguyen, T.V. Tran, L.X. Nong, T.T. Nguyen, L.G. Bach, B. Abdullah, S.S. Hong, T.V. Nguyen. 2020a. $BiVO_4$ photocatalysis design and applications to hydrogen production, degradation of organic compounds and reduction of CO_2 to fuels: a review. *Environmental Chemistry Letters*. Doi: 10.1007/s10311-020-01039-0.

Nguyen, T.D., T.V. Tran, S. Singh, P.T.T. Phuong, L.G. Bach, S. Nanda, D.V.N. Vo. 2020b. Conversion of carbon dioxide into formaldehyde. In: *Conversion of Carbon Dioxide into Hydrocarbons. Vol. 2 Technology*; eds. A.M. Asiri, Inamuddin, L. Eric; 159–183. Singapore: Springer Nature.

Okolie, J.A., S. Nanda, A.K. Dalai, F. Berruti, J.A. Kozinski. 2020a. A review on subcritical and supercritical water gasification of biogenic, polymeric and petroleum wastes to hydrogen-rich synthesis gas. *Renewable and Sustainable Energy Reviews* 119:109546.

Okolie, J.A., S. Nanda, A.K. Dalai, J.A. Kozinski. 2020b. Chemistry and specialty industrial applications of lignocellulosic biomass. *Waste and Biomass Valorization*. Doi: 10.1007/s12649-020-01123-0.

Okolie, J.A., S. Nanda, A.K. Dalai, J.A. Kozinski. 2020c. Hydrothermal gasification of soybean straw and flax straw for hydrogen-rich syngas production: experimental and thermodynamic modeling. *Energy Conversion and Management* 208:112545.

Onoghwarite, O.E., N.V. Obiora, E.A. Ben, N.O.E. Moses. 2016. Bioethanol production from corn stover using saccharomyces cerevisiae. *International Journal of Scientific & Engineering Research* 7:290–293.

Parakh, P.D., S. Nanda, J.A. Kozinski. 2020. Eco-friendly transformation of waste biomass to biofuels. *Current Biochemical Engineering* 6:120–134.

Patil, P.N., P.R. Gogate, L. Csoka, A. Dregelyi-Kiss, M. Horvath. 2016. Intensification of biogas production using pretreatment based on hydrodynamic cavitation. *Ultrasonic Sonochemistry* 30:79–86.

Peral, C. 2016. Biomass pretreatment strategies (technologies, environmental performance, economic considerations, industrial implementation). In: *Biotransformation of Agricultural Waste and By-Products: The Food, Feed, Fibre, Fuel (4F) Economy*; ed. P. Poltronieri, O.F. D'Urso; 125–160. Elsevier.

Pielhop, T., J. Amgarten, P.R. von Rohr, M.H. Studer. 2016. Steam explosion pretreatment of softwood: the effect of the explosive decompression on enzymatic digestibility. *Biotechnology for Biofuels* 9:152.

Rana, R., S. Nanda. 2019. Application of methanation as an alternate energy technology. In: *Biogas Technology*; eds. S. Mishra, T.K. Adhya, S.K. Ojha; 105–116. New Delhi, India: New India Publishing Agency.

Rana, R., S. Nanda, A. Maclennan, Y. Hu, J.A. Kozinski, A.K. Dalai. 2019a. Comparative evaluation for catalytic gasification of petroleum coke and asphaltene in subcritical and supercritical water. *Journal of Energy Chemistry* 31:107–118.

Rana, R., S. Nanda, A.K. Dalai, J.A. Kozinski, J. Adjaye. 2019b. Synthetic crude processing: impacts of fine particles on hydrotreating of bitumen-derived gas oil. In: *Fuel Processing and Energy Utilization*; eds. S. Nanda, P.K. Sarangi, D.V.N. Vo; 187–206. Boca Raton, FL: CRC Press.

Rana, R., S. Nanda, J.A. Kozinski, A.K. Dalai. 2018a. Investigating the applicability of Athabasca bitumen as a feedstock for hydrogen production through catalytic supercritical water gasification. *Journal of Environmental Chemical Engineering* 6:182–189.

Rana, R., S. Nanda, S.N. Reddy, A.K. Dalai, J.A. Kozinski, I. Gökalp. 2020. Catalytic gasification of light and heavy gas oils in supercritical water. *Journal of the Energy Institute*. 93:2025-2032.

Rana, R., S. Nanda, V. Meda, A.K. Dalai, J.A. Kozinski. 2018b. A review of lignin chemistry and its biorefining conversion technologies. *Journal of Biochemical Engineering and Bioprocess Technology* 1:2.

Reddy, S.N., S. Nanda, A.K. Dalai, J.A. Kozinski. 2014. Supercritical water gasification of biomass for hydrogen production. *International Journal of Hydrogen Energy* 39:6912–6926.

Reddy, S.N., S. Nanda, J.A. Kozinski. 2016. Supercritical water gasification of glycerol and methanol mixtures as model waste residues from biodiesel refinery. *Chemical Engineering Research and Design* 113:17–27.

Reddy, S.N., S. Nanda, P. Kumar, M.C. Hicks, U.G. Hegde, J.A. Kozinski. 2019. Impacts of oxidant characteristics on the ignition of n-propanol-air hydrothermal flames in supercritical water. *Combustion and Flame* 203:46–55.

Reddy, S.N., S. Nanda, P.K. Sarangi. 2018. Applications of supercritical fluids for biodiesel production. In: *Recent Advancements in Biofuels and Bioenergy Utilization*; eds. P.K. Sarangi, S. Nanda, P. Mohanty; 261–284. Singapore: Springer Nature.

Reddy, S.N., S. Nanda, U.G. Hegde, M.C. Hicks, J.A. Kozinski. 2015. Ignition of hydrothermal flames. *RSC Advances* 5:36404–36422.

Reddy, S.N., S. Nanda, U.G. Hegde, M.C. Hicks, J.A. Kozinski. 2017. Ignition of n-propanol–air hydrothermal flames during supercritical water oxidation. *Proceedings of the Combustion Institute* 36:2503–2511.

Ruiz, H.A., R.M. Rodríguez-Jasso, B.D. Fernandes, A.A. Vicente, J.A. Teixeira. 2013. Hydrothermal processing, as an alternative for upgrading agriculture residues and marine biomass according to the biorefinery concept: a review. *Renewable and Sustainable Energy Reviews* 21:35–51.

Saharan, V.K., R. Manava, M. Aqeel, A.B. Pandit. 2013. Effect of geometry of hydrodynamically cavitating device on degradation of orange-G. *Ultrasonic Sonochemistry* 20:345–353.

Sarangi, P.K., S. Nanda. 2018. Recent developments and challenges of acetone-butanol-ethanol fermentation. In: *Recent Advancements in Biofuels and Bioenergy Utilization*; eds. P.K. Sarangi, S. Nanda, P. Mohanty; 111–123. Singapore: Springer Nature.

Sarangi, P.K., S. Nanda. 2019a. Bioconversion of agro-wastes into phenolic flavour compounds. In: *Biotechnology for Sustainable Energy and Products*; eds. P.K. Sarangi, S. Nanda; 266–284. New Delhi: I.K. International Publishing House Pvt. Ltd.

Sarangi, P.K., S. Nanda. 2019b. Recent advances in consolidated bioprocessing for microbe-assisted biofuel production. In: *Fuel Processing and Energy Utilization*; eds. S. Nanda, P.K. Sarangi, D.V.N. Vo; 141–157. Boca Raton, FL: CRC Press.

Sarangi, P.K., S. Nanda. 2020. Biohydrogen production through dark fermentation. *Chemical Engineering & Technology* 43:601–612.

Sarangi, P.K., S. Nanda, D.V.N. Vo. 2020. Technological advancements in the production and application of biomethanol. In: *Biorefinery of Alternative Resources: Targeting Green Fuels and Platform Chemicals*; eds. S. Nanda, D.V.N. Vo, P.K. Sarangi; 127–139. Singapore: Springer Nature.

Shafiqah, M.N.N., H.N. Tran, T.D. Nguyen, P.T.T. Phuong, B. Abdullah, S.S. Lam, P. Nguyen-Tri, R. Kumar, S. Nanda, D.V.N. Vo. 2020. Ethanol CO_2 reforming on La_2O_3 and CeO_2-promoted Cu/Al_2O_3 catalysts for enhanced hydrogen production. *International Journal of Hydrogen Energy* 45:18398–18410.

Sills, D.L., J.M. Gossett. 2011. Assessment of commercial hemicellulases for saccharification of alkaline pretreated perennial biomass. *Bioresource Technology* 102:1389–1398.

Sindhu, R., P. Binod, A. Pandey. 2016. Biological pretreatment of lignocellulosic biomass: an overview. *Bioresource Technology* 199:76–82.

Singh, A., S. Nanda, F. Berruti. 2020. A review of thermochemical and biochemical conversion of miscanthus to biofuels. In: *Biorefinery of Alternative Resources: Targeting Green Fuels and Platform Chemicals*; eds. S. Nanda, D.V.N. Vo, P.K. Sarangi; 195–220. Singapore: Springer Nature.

Singh, S., R. Kumar, H.D. Setiabudi, S. Nanda, D.V.N. Vo. 2018. Advanced synthesis strategies of mesoporous SBA-15 supported catalysts for catalytic reforming applications: a state-of-the-art review. *Applied Catalysis A: General* 559:57–74.

Sun, J., L. Xu, G.H. Dong, S. Nanda, H. Li, Z. Fang, J.A. Kozinski, A.K. Dalai. 2020. Subcritical water gasification of lignocellulosic wastes for hydrogen production with Co modified Ni/Al_2O_3 catalyst. *The Journal of Supercritical Fluids* 162:104863.

Valdivia, M., J.L. Galan, J. Laffarga, J.L. Ramos. 2016. Biofuels 2020: biorefineries based on lignocellulosic materials. *Microbial Biotechnology* 9:585–594.

Vivekanand, V., E.F. Olsen, V.G.H. Eijsink, S.J. Horn. 2013. Effect of different steam explosion conditions on methane potential and enzymatic saccharification of birch. *Bioresource Technology* 127:343–349.

Xu, C., D. Singh, K.M. Dorgan, X. Zhang, S. Chen. 2015. Screening of ligninolytic fungi for biological pretreatment of lignocellulosic biomass. *Canadian Journal of Microbiology* 61:745–752.

Yadav, P., S.N. Reddy, S. Nanda. 2019. Cultivation and conversion of algae for wastewater treatment and biofuel production. In: *Fuel Processing and Energy Utilization*; eds. S. Nanda, P.K. Sarangi, D.V.N. Vo; 159–175. Boca Raton, FL: CRC Press.

Yang, G., J. Wang. 2019. Ultrasound combined with dilute acid pretreatment of grass for improvement of fermentative hydrogen production. *Bioresource Technology* 275:10–18.

Yoo, C.G., Y. Pu, A.J. Ragauskas. 2017. Ionic liquids: promising green solvents for lignocellulosic biomass utilization. *Current Opinion in Green Sustainable Chemistry* 5:5–11.

Yu, G., S. Yano, H. Inoue, S. Inoue, T. Endo, S. Sawayama. 2010. Pretreatment of rice straw by a hot-compressed water process for enzymatic hydrolysis. *Applied Biochemistry Biotechnology* 160:539–551.

Yu, Z., B. Zhang, F. Yu, G. Xu, A. Song. 2012. A real explosion: the requirement of steam explosion pretreatment. *Bioresource Technology* 121:335–341.

Zhu, J.Y., X.J. Pan, G.S. Wang, R. Gleisner. 2009. Sulfite pretreatment for robust enzymatic saccharification of spruce and red pine. *Bioresource Technology* 100:2411–2418.

Bioconversion of Waste Biomass to Bioethanol

3

3.1 INTRODUCTION

The concept of biorefinery is an outcome of advancements in biotechnology, chemical engineering, bioresource technology, process chemistry, genetic engineering, industrial engineering and other cross-disciplinary areas, all of which support research and development on the conversion of alternative resources to value-added fuels, commodity chemicals and industrially relevant products. The production of biofuels and biochemicals from various waste biomass has gained immense interest in research and applications in the past few decades as far as the bio-based economy and circular economy are concerned (Nanda et al. 2015b; Sarangi and Nanda 2020). Renewable organic wastes including lignocellulosic feedstocks (e.g. agricultural biomass, forestry residues and energy crops), microalgae and other biogenic wastes (e.g. municipal solid waste, food waste, waste cooking oil, cattle manure, sewage sludge and industrial effluents) have huge potentials to produce biofuels through biological and thermochemical conversion for supplementing the global energy requirements (Nanda et al. 2016a; Nanda et al. 2016b; Nanda et al. 2016c; Nanda et al. 2016e; Reddy et al. 2016; Gong et al. 2017a; Gong et al. 2017b; Nanda et al. 2017c; Nanda et al. 2018b; Nanda et al. 2019). The biofuels produced from the above-mentioned lignocellulosic biomass and biogenic wastes are termed as second-generation biofuels (Nanda et al. 2018a).

Bioethanol is one of the second-generation biofuels and base chemicals produced from the biological conversion of lignocellulosic biomass and organic wastes. Bioethanol can be blended with petrol (or gasoline) in various proportions such as E10, E85 and E95 containing 10%, 85% and 95% of ethanol, respectively. The blending of ethanol at lower proportions with gasoline is preferred for use in the existing vehicular engines since higher proportion necessitates significant mechanical modifications to the automobile engines (Sarangi and Nanda 2018). Moreover, ethanol has an oxygen content of 35% and is completely soluble in water at 25°C, which causes technical issues in higher blending ratios with gasoline. The calorific value of ethanol (C_2H_5OH) is 21.2 MJ/kg while that of gasoline is 32.5 MJ/kg (Nanda et al. 2017b). Similarly, a few other fuel properties such as research octane number, motor octane number and the air-fuel ratio of ethanol are 129, 102 and 9, respectively (Nanda et al. 2017b).

This chapter gives an overview of bioethanol production from waste lignocellulosic biomass. The chapter describes the potential of waste biomass for bioethanol production. It also provides insights on microbial fermentation for bioethanol production along with the biomass pretreatment and bioprocess parameters. The chapter concludes with a note on technical challenges and future possibilities.

3.2 POTENTIAL OF LIGNOCELLULOSIC BIOMASS

When compared to fossil fuels, waste plant biomass is considered to be economical, abundant and renewable, making them one of the most prominent and trusted sources of renewable energy (Sarangi and Nanda 2019a; Sarangi and Nanda 2019b; Sarangi and Nanda 2019c). Bioethanol was traditionally produced from first-generation feedstocks mostly comprising food crops and grains (e.g., maize, potatoes, wheat, cassava and other starch-based crops). However, the massive diversion of these food crops to biorefineries for bioethanol production contributed to a shortage in the food supply, rising food prices, the competition to cultivable lands and the socio-environmental unrest relating to the "food *versus* fuel" controversy (Nanda et al. 2015b).

With the unpopularity of first-generation bioethanol, soon emerged the second-generation bioethanol produced from inedible plant residues that have no competition to food supply and arable lands. These inedible plant residues are mostly the lignocellulosic biomass. The accessibility of biomass resources and their potential to produce biofuel should be focused to fulfill the demand of bioethanol. Food security should not be compromised by any nation while addressing domestic energy security. Hence, it becomes an important consideration to resolve the energy crisis along with the suitable valorization of agricultural and forestry refuse as well as organic wastes economically and sustainably.

Cellulose constitutes about 35–55 wt% of lignocellulosic biomass followed by hemicelluloses (20–40 wt%) and lignin (10–25 wt%) (Nanda et al. 2013). Trace presence of inorganic ingredients (i.e. mineral matter and ash), nitrogenous compounds, waxes and extractives (e.g. pectin, ester, ether, resins, tannins, terpenoids, chlorophyll and other polar and non-polar components) is also found in lignocellulosic biomass (Okolie et al. 2019). Hydrogen and covalent bonds link cellulose and hemicellulose firmly with lignin, which makes the lignocellulosic framework recalcitrant to chemical agents and enzymes. Straight chains of D-glucose subunits are found in cellulose connected with β-(1,4)-glycosidic bonds. Cellulose occurs in both amorphous and crystalline nature in biomass (Nanda et al. 2016d). Hemicelluloses are matrix polysaccharides containing both hexose sugars (e.g. glucose, rhamnose, galactose and mannose) and pentose sugars (e.g. xylose and arabinose) as well as sugar acids (e.g. glucuronic acid and galacturonic acid) (Okolie et al. 2020). Lignin is a highly branched cross-linked aromatic phenylpropane polymer with hydrophobic properties (Fougere et al. 2016). It is synthesized from phenylpropanoid precursors that lead to the synthesis of these polyphenolic aromatic compounds. The basic building blocks of lignin are p-coumaryl alcohol, coniferyl alcohol and sinapyl alcohol (Rana et al. 2018).

3.3 UPSTREAM AND DOWNSTREAM TECHNOLOGIES IN BIOETHANOL PRODUCTION

The production of second-generation ethanol is typically a four-stage process such as: (i) biomass pretreatment and hydrolysis, (ii) enzymatic saccharification, (iii) microbial fermentation and (iv) product separation (Nanda et al. 2014a; Sarangi et al. 2020). A suitable physicochemical, chemimechanical or hydrothermal pretreatment causes structural modifications to lignocellulosic biomass, thus reducing cellulose crystallinity, depolymerizing lignin and separating hemicellulose (Nanda et al. 2014b). A hydrolytic pretreatment of biomass enhances the efficiency of subsequent enzyme-mediated saccharification of fermentable sugars (pentose and hexose) recovered from cellulose and hemicellulose. Enzymatic saccharification enhances the near-complete release of monomeric sugars from the hydrolyzed carbohydrates and polysaccharides in the biomass for microbial fermentation.

Several cellulolytic enzymes (e.g. cellulases), hemicellulases, lignin-modifying and lignin-degrading enzymes are involved in enzymatic saccharification and hydrolysis of lignocellulosic biomass (Parakh et al. 2020). In the next step, suitable microorganisms (individual or consortium) are used to ferment the monomeric sugars recovered from biomass because of pretreatment and enzymatic saccharification to the desired alcohol and other organic compounds (e.g., ethanol, butanol, methanol, acetone, etc.) (Sarangi and Nanda 2019b). In the final downstream stage, the products are separated from the fermentation media and purified based on their chemical and solvent properties through a wide variety of technologies such as distillation, gas stripping, liquid-liquid separation, adsorption, perstraction, pervaporation, supercritical CO_2 fractionation, etc. (Nanda et al. 2017a). The typical upstream and downstream process involved in the production of bioethanol from lignocellulosic biomass is illustrated in Figure 3.1.

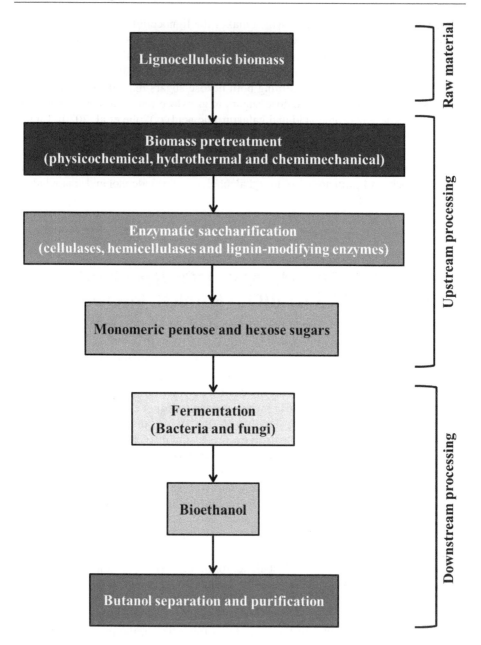

FIGURE 3.1   Typical upstream and downstream steps involved in lignocellulosic biomass conversion to bioethanol

3.4 ENZYMATIC SACCHARIFICATION

Enzymatic hydrolysis plays a key role in determining the operating costs involved in the production of second-generation bioethanol. An efficient enzymatic saccharification could lead to the maximum recovery of fermentable monomeric sugars from the biomass, thus leading to an improved fermentation for bioethanol production. The first-generation biomass, which predominantly contains starch, is relatively easier to hydrolyze and ferment for bioethanol production. In contrast, lignocellulosic biomass contains highly robust and biochemically stable lignin, which creates hindrances in the access of enzymes and other pretreatment agents for denaturing the cellulose-hemicellulose-lignin matrix and releasing the sugars (Nanda et al. 2015a).

Cellulases (i.e. cellulose-degrading enzymes) along with hemicellulase (i.e. hemicellulose-degrading enzymes) are utilized for a complete breakdown of all polysaccharides in lignocellulosic biomass into sugar monomers. Cellulases are a pivotal and prominent constituent of the enzyme cocktail used for saccharification because of the higher degree of polymerization of cellulose and their crystallinity than that of hemicellulose (Sukharnikov et al. 2011). Van der Waals interactions along with hydrogen bonds prevailing in between the glucose monomers are responsible for rendering recalcitrance to cellulose fibers. Therefore, reducing cellulose crystallinity and increasing amorphous cellulose moieties are key considerations for enzymatic saccharification. With specific functionalities of degrading cellulose, the types of cellulase enzymes are endocellulases, exocellulases, cellobiases (or β-glucosidases), oxidative cellulases and cellulose phosphorylases.

A variety of glycoside hydrolases are used in enzymatic saccharification of complex lignocelluloses. The family of glycoside hydrolases includes cellulases, hemicellulases, pectin-degrading enzymes and lignin-degrading enzymes. More than 130 glycoside hydrolases families have been explored for the conversion of complex carbohydrates into simpler sugars (Lombard et al. 2014), out of which 40 are cellulolytic enzymes with the ability to achieve high-efficiency cellulose hydrolysis with well-coordinated synergy for bioethanol production (Liu et al. 2018).

Cellulases, hemicellulases and β-glucosidases are the main constituents of the cocktail that manifest the conversion of polysaccharides into pentose and hexose monomers. The second-generation bioethanol production can be made cost-effective by engineering the bioprocess to recycle the expensive enzyme cocktail, which significantly adds to the process expenditures. Strategies should also be focused on enhancing the efficiency of enzymes either by process engineering or by upstream modifications (e.g. efficient biomass pretreatment with maximum hydrolysis). Another strategy is genetic engineering, which involves high-performance mutagenesis and engineered energy crops. Second, maximum hydrolysis can be achieved if an enzyme cocktail can act best under specified process conditions while retaining its thermal stability and viability. Enzymes recovered from extremophilic microorganisms could be a possible solution (Miller and Blum 2010).

Thermophilic enzymes can make bioprocess technology cost-effective as they enhance the overall performance of the process by preventing contamination of hydrolysates. In such a scenario, saccharification can perform at temperatures relatively higher than that preferred by contaminating microorganisms. Under ambient conditions, soil microbial heterotrophs are supported by their efficiencies to undergo enzymatic hydrolysis of lignocellulosic materials through natural decomposition, which is also an important process for carbon cycling in terrestrial ecosystems (Nanda et al. 2016f).

3.5 MICROBIAL FERMENTATION

Several microorganisms including fungi (e.g. *Aspergillus*, *Candida shehatae*, *Fusarium* sp., *Kluyveromyces* sp., *Neurospora* sp., *Phanerochaete* sp., *Penicillium* sp., *Pichia kudriavzevii*, *Saccharomyces cerevisiae*, *Schizophyllum* sp., *Sclerotium* sp., *Trichoderma* sp., etc.) and bacteria (e.g. *Acetovibrio* sp., *Bacillus* sp., *Clostridium thermocellum*, *Erwinia* sp., *Escherichia coli*, *Klebsiella oxytoca*, *Ruminococcus* sp., *Zymomonas mobilis*, etc.) accomplish fermentation of biomass hydrolysates to produce ethanol (Nanda et al. 2014b). *S. cerevisiae* is a model microorganism for ethanol fermentation because of its high efficiency, stability, a faster rate of sugar conversion and high solvent (alcohol) tolerance. Moreover, it is also considered as GRAS (generally regarded as safe). *S. cerevisiae* is also a potential producer of zymase, an enzyme complex that manifests the biocatalysis of sugar fermentation into ethanol and CO_2 (Lin and Tanaka 2006).

Although *S. cerevisiae* is efficient in fermenting hexose sugars (glucose) and starch, it lacks the natural ability to ferment pentose sugars (i.e. xylose and arabinose) from hemicelluloses. However, a significant development in synthetic biology and genetic engineering has made it possible to express metabolically engineered pathways for D-xylose and L-arabinose metabolism in *S. cerevisiae* (Nijland and Driessen 2020). Moreover, a few fungi such as *Candida parapsilosis*, *Candida shehatae* and *Pichia stipitis* have demonstrated the xylose metabolism with the aid of xylose reductase and xylitol dehydrogenase (Nanda et al. 2014b). Xylose reductase transforms xylose to xylitol, whereas xylitol dehydrogenase further converts xylitol to xylulose. Xylulose can be metabolized through the pentose-phosphate pathway.

Pretreatment and enzymatic saccharification generate simple monomeric sugars, which serve as the source of carbon and energy for various microorganisms. Fermentation typically operates in the temperature range of 30–36°C, whereas the process of enzymatic hydrolysis demands a temperature range of 45–50°C. The two most widely used fermentation techniques are solid-state fermentation and submerged fermentation. Solid-state fermentation technology employs the microbial growth on moist solid substrates for producing high value-added products. This technology is an expanding approach for the commercial production of many enzymes. This technology has grabbed enormous attention since it is better than submerged fermentation, which involved fermentation in a liquid-based (i.e. hydrolysate-rich) media.

Another iteration of fermentation, i.e. separate hydrolysis and fermentation (SHF) is easy to manage the enzymatic hydrolysis and fermentation processes. However, SHF encounters a limitation of accumulation of glucose and cellobiose during enzymatic hydrolysis, which may cause feedback inhibition for the hydrolytic enzymes. To resolve such an issue, the addition of β-glucosidase becomes necessary, thereby making the process costly. The necessity of β-glucosidase can be curbed in simultaneous saccharification and fermentation (SSF) process due to restricted chances of feedback inhibition, which can make SSF cost-effective than SHF (Elumalia and Thangavelu 2010; Kont et al. 2013). Furthermore, SSF is a faster process with a higher yield of ethanol in contrast to SHF. The presence of ethanol in the fermentation medium creates an environment where the chances of contamination and spoilage of the sugar-rich hydrolysate are less (Sasikumar and Viruthagiri 2010).

3.6 CONCLUSIONS

The production of second-generation biofuels such as bioethanol from lignocellulosic biomass plays a pivotal function in the sustainability of biorefineries. As the building blocks of these renewable feedstocks, recovery of the fermentable sugars is imperative for a highly efficient and profitable bioethanol production process. Genetically modified microorganisms and their engineered metabolisms can help utilize a wide variety of monomeric sugars extracted from the pretreatment and enzymatic hydrolysis of lignocellulosic biomass. Furthermore, optimization of the fermentation process and minimizing the chances of contamination and undesired byproduct formation can lead to improve bioethanol yields and lower the overall process expenditures. Pretreatment and enzymatic hydrolysis are crucial steps for the recovery of fermentable sugars from heterogeneous lignocellulosic biomass. Novel pretreatment processes can be adopted that can allow enzymes to act upon biomass simultaneously, thereby maximizing the extraction of monomeric sugars.

REFERENCES

Elumalia, S., V. Thangavelu. 2010. Simultaneous saccharification and fermentation (SSF) of pretreated sugarcane bagasse using cellulose and *Saccharomyces cerevisiae*-kinetics and modeling. *Chemical Engineering Research Bulletin* 14:29–35.
Fougere, D., S. Nanda, K. Clarke, J.A. Kozinski, K. Li. 2016. Effect of acidic pretreatment on the chemistry and distribution of lignin in aspen wood and wheat straw substrates. *Biomass & Bioenergy* 91:56–68.
Gong, M., S. Nanda, H.N. Hunter, W. Zhu, A.K. Dalai, J.A. Kozinski. 2017a. Lewis acid catalyzed gasification of humic acid in supercritical water. *Catalysis Today* 291:13–23.

Gong, M., S. Nanda, M.J. Romero, W. Zhu, J.A. Kozinski. 2017b. Subcritical and supercritical water gasification of humic acid as a model compound of humic substances in sewage sludge. *The Journal of Supercritical Fluids* 119:130–138.

Kont, R., M. Kurasin, H. Teugjas, P. Valjamae. 2013. Strong cellulase inhibitors from the hydrothermal pretreatment of wheat straw. *Biotechnology for Biofuels* 6:135.

Lin, Y., S. Tanaka. 2006. Ethanol fermentation from biomass resources: current state and prospects. *Applied Microbiology and Biotechnology* 69:627–642.

Liu, H., B. Pang, Y. Zhao, J. Lua, Y. Han, H. Wang. 2018. Comparative study of two different alkali-mechanical pretreatments of corn stover for bioethanol production. *Fuel* 221:21–27.

Lombard, V., H.G. Ramulu, E. Drula, P.M. Coutinho, B. Henrissat. 2014. The carbohydrate-active enzymes database (CAZy) in 2013. *Nucleic Acids Research* 42:D490–D495.

Miller, P.S., P.H. Blum. 2010. Extremophile-inspired strategies for enzymatic biomass saccharification. *Environmental Technology* 31:1005–1015.

Nanda, S., A.K. Dalai, I. Gökalp, J.A. Kozinski. 2016a. Valorization of horse manure through catalytic supercritical water gasification. *Waste Management* 52:147–158.

Nanda, S., A.K. Dalai, J.A. Kozinski. 2014a. Butanol and ethanol production from lignocellulosic feedstock: biomass pretreatment and bioconversion. *Energy Science & Engineering* 2:138–148.

Nanda, S., A.K. Dalai, J.A. Kozinski. 2016b. Supercritical water gasification of timothy grass as an energy crop in the presence of alkali carbonate and hydroxide catalysts. *Biomass & Bioenergy* 95:378–387.

Nanda, S., A.K. Dalai, J.A. Kozinski. 2017a. Butanol from renewable biomass: highlights on downstream processing and recovery techniques. In: *Sustainable Utilization of Natural Resources*; eds. P. Mondal, A.K. Dalai; 187–211. Boca Raton, FL: CRC Press.

Nanda, S., D. Golemi-Kotra, J.C. McDermott, A.K. Dalai, I. Gökalp, J.A. Kozinski. 2017b. Fermentative production of butanol: perspectives on synthetic biology. *New Biotechnology* 37:210–221.

Nanda, S., J. Isen, A.K. Dalai, J.A. Kozinski. 2016c. Gasification of fruit wastes and agro-food residues in supercritical water. *Energy Conversion and Management* 110:296–306.

Nanda, S., J. Maley, J.A. Kozinski, A.K. Dalai. 2015a. Physico-chemical evolution in lignocellulosic feedstocks during hydrothermal pretreatment and delignification. *Journal of Biobased Materials and Bioenergy* 9:295–308.

Nanda, S., J. Mohammad, S.N. Reddy, J.A. Kozinski, A.K. Dalai. 2014b. Pathways of lignocellulosic biomass conversion to renewable fuels. *Biomass Conversion and Biorefinery* 4:157–191.

Nanda, S., J.A. Kozinski, A.K. Dalai. 2016d. Lignocellulosic biomass: a review of conversion technologies and fuel products. *Current Biochemical Engineering* 3:24–36.

Nanda, S., M. Gong, H.N. Hunter, A.K. Dalai, I. Gökalp, J.A. Kozinski. 2017c. An assessment of pinecone gasification in subcritical, near-critical and supercritical water. *Fuel Processing Technology* 168:84–96.

Nanda, S., P. Mohanty, K.K. Pant, S. Naik, J.A. Kozinski, A.K. Dalai. 2013. Characterization of North American lignocellulosic biomass and biochars in terms of their candidacy for alternate renewable fuels. *Bioenergy Research* 6:663–677.

Nanda, S., R. Azargohar, A.K. Dalai, J.A. Kozinski. 2015b. An assessment on the sustainability of lignocellulosic biomass for biorefining. *Renewable and Sustainable Energy Reviews* 50:925–941.

Nanda, S., R. Rana, H.N. Hunter, Z. Fang, A.K. Dalai, J.A. Kozinski. 2019. Hydrothermal catalytic processing of waste cooking oil for hydrogen-rich syngas production. *Chemical Engineering Science* 195:935–945.

Nanda, S., R. Rana, P.K. Sarangi, A.K. Dalai, J.A. Kozinski. 2018a. A broad introduction to first, second and third generation biofuels. In: *Recent Advancements in Biofuels and Bioenergy Utilization*; eds. P.K. Sarangi, S. Nanda, P. Mohanty; 1–25. Singapore: Springer Nature.

Nanda, S., S.N. Reddy, A.K. Dalai, J.A. Kozinski. 2016e. Subcritical and supercritical water gasification of lignocellulosic biomass impregnated with nickel nanocatalyst for hydrogen production. *International Journal of Hydrogen Energy* 41:4907–4921.

Nanda, S., S.N. Reddy, D.V.N. Vo, B.N. Sahoo, J.A. Kozinski. 2018b. Catalytic gasification of wheat straw in hot compressed (subcritical and supercritical) water for hydrogen production. *Energy Science & Engineering* 6:448–459.

Nanda, S., S.N. Reddy, S.K. Mitra, J.A. Kozinski. 2016f. The progressive routes for carbon capture and sequestration. *Energy Science & Engineering* 4:99–122.

Nijland, J.G., A.J.M. Driessen. 2020. Engineering of pentose transport in *saccharomyces cerevisiae* for biotechnological applications. *Frontiers in Bioengineering and Biotechnology* 7:464.

Okolie, J.A., R. Rana, S. Nanda, A.K. Dalai, J.A. Kozinski. 2019. Supercritical water gasification of biomass: a state-of-the-art review of process parameters, reaction mechanisms and catalysis. *Sustainable Energy & Fuels* 3:578–598.

Okolie, J.A., S. Nanda, A.K. Dalai, J.A. Kozinski. 2020. Chemistry and specialty industrial applications of lignocellulosic biomass. *Waste and Biomass Valorization*. Doi: 10.1007/s12649-020-01123-0.

Parakh, P.D., S. Nanda, J.A. Kozinski. 2020. Eco-friendly transformation of waste biomass to biofuels. *Current Biochemical Engineering* 6:120–134.

Rana, R., S. Nanda, V. Meda, A.K. Dalai, J.A. Kozinski. 2018. A review of lignin chemistry and its biorefining conversion technologies. *Journal of Biochemical Engineering and Bioprocess Technology* 1:2.

Reddy, S.N., S. Nanda, J.A. Kozinski. 2016. Supercritical water gasification of glycerol and methanol mixtures as model waste residues from biodiesel refinery. *Chemical Engineering Research and Design* 113:17–27.

Sarangi, P.K., S. Nanda. 2018. Recent developments and challenges of acetone-butanol-ethanol fermentation. In: *Recent Advancements in Biofuels and Bioenergy Utilization*; eds. P.K. Sarangi, S. Nanda, P. Mohanty; 111–123. Singapore: Springer Nature.

Sarangi, P.K., S. Nanda. 2019a. Bioconversion of agro-wastes into phenolic flavour compounds. In: *Biotechnology for Sustainable Energy and Products*; eds. P.K. Sarangi, S. Nanda; 266–284. New Delhi: I.K. International Publishing House Pvt. Ltd.

Sarangi, P.K., S. Nanda. 2019b. Recent advances in consolidated bioprocessing for microbe-assisted biofuel production. In: *Fuel Processing and Energy Utilization*; eds. S. Nanda, P.K. Sarangi, D.V.N. Vo; 141–157. Boca Raton, FL: CRC Press.

Sarangi, P.K., S. Nanda. 2019c. Valorization of pineapple wastes for biomethane generation. In: *Biogas Technology*; eds. S. Mishra, T.K. Adhya, S.K. Ojha; 169–180. New Delhi, India: New India Publishing Agency.

Sarangi, P.K., S. Nanda. 2020. Biohydrogen production through dark fermentation. *Chemical Engineering & Technology* 43:601–612.

Sarangi, P.K., S. Nanda, D.V.N. Vo. 2020. Technological advancements in the production and application of biomethanol. In: *Biorefinery of Alternative Resources: Targeting Green Fuels and Platform Chemicals*; eds. S. Nanda, D.V.N. Vo, P.K. Sarangi; 127–139. Singapore: Springer Nature.

Sasikumar, E., T. Viruthagiri. 2010. Simultaneous saccharification and fermentation (SSF) of sugarcane bagasse - kinetics and modeling. *International Journal of Chemical and Biological Engineering* 3:57–64.

Sukharnikov, L.O., B.J. Cantwell, M. Podar, I.B. Zhulin. 2011. Cellulases: ambiguous non-homologous enzymes in a genomic perspective. *Trends in Biotechnology* 29:473–479.

Bioconversion of Waste Biomass to Biobutanol

4

4.1 INTRODUCTION

There has been a growing interest in the production and utilization of synthetic renewable transportation fuels due to rising crude oil prices, mounting demand for fossil fuels, the adverse impact of greenhouse gas emissions and the resulting global warming (Nanda et al. 2015b; Nanda et al. 2016a; Nanda et al. 2016g; Nanda et al. 2017d). Hence, the worldwide interest in deploying biofuels and biochemicals in addition to their production from renewable biomass and wastes is gaining momentum (Sarangi and Nanda 2018; Sarangi and Nanda 2019a; Sarangi and Nanda 2019b; Sarangi and Nanda 2020; Sarangi et al. 2020). First-generation biofuels are criticized severely as far as their sustainability and competition to food supply chain and cultivable lands are concerned (Nanda et al. 2018a). On the other hand, next-generation biofuels provide a more sustainable platform ensuring both energy security and food security as their production relies on mostly inedible plant residues including agricultural crop refuse (Nanda et al. 2018b; Sun et al. 2020; Okolie et al. 2020c), forestry biomass (Nanda et al. 2016f; Nanda et al. 2017c), dedicated energy crops (Nanda et al. 2016c; Singh et al. 2020), cattle manure (Nanda et al. 2016b), municipal solid wastes (Okolie et al. 2020a), food waste (Nanda et al. 2015c; Nanda et al. 2016d; Nanda et al. 2019a), industrial effluents (Nanda et al. 2015d), sewage sludge (Gong et al. 2017a; Gong et al. 2017b), polymeric wastes (Nanda et al. 2019b) and petroleum residues (Rana et al. 2018a; Rana et al. 2019; Rana et al. 2020).

Lignocellulosic biomass, consisting of agricultural crop residues, dedicated energy crops, invasive plants and forestry biomass, is a storehouse of renewable natural polymers (i.e. lignin) and polysaccharides (i.e. cellulose and hemicellulose) that can be converted through thermochemical and biological technologies to solid (e.g. biochar, torrefied biomass and fuel pellets), liquid (e.g. bio-oil, bioethanol, biobutanol, biodiesel, etc.) and gaseous biofuels (e.g. biohydrogen, biomethane, syngas, etc.) (Nanda et al., 2016e; Azargohar et al. 2019; Kang et al. 2020; Parakh et al. 2020; Okolie et al. 2020b).

Hence, there are different alternatives to seek future biofuel solutions for the transportation sectors through the utilization of next-generation bioenergy feedstocks.

In contrast to bioethanol (C_2H_5OH), biobutanol (C_4H_9OH) appears to be a superior fuel with advanced properties, a few of which are less corrosiveness, higher calorific value (29.2 MJ/L), lower volatility and less hygroscopic (7.3% soluble in water), gasoline-equivalent research octane number (96), motor octane number (78) and air-to-fuel ratio (11.2), lower oxygen content (22%), less flammability and reduced hazardousness for handling (Nanda et al. 2014a; Nanda et al. 2017a; Nanda et al. 2017b; Sarangi and Nanda 2018; Nanda et al. 2020). Due to its lower vapor pressure and less hygroscopic nature, biobutanol can be transported in the gasoline supply chain pipelines to the fueling stations even in the cold weather (Qureshi and Ezeji 2008). Owing to its fuel properties similar to gasoline, biobutanol can either be blended with gasoline in flexible ratios or be used as a drop-in fuel without blending in the current gasoline-fueled automobile engines. Biobutanol can be produced biologically through the traditional acetone-butanol-ethanol (ABE) fermentation, although there are a few technical bottlenecks associated with its fermentative production such as lower product yields, bacteriophage contamination, product separation, expensive process, etc. This chapter discusses a few of such attributes in the biobutanol production through ABE fermentation from lignocellulosic biomass.

4.2 PRETREATMENT OF LIGNOCELLULOSIC BIOMASS

Although lignocellulosic biomasses are found to be promising next-generation bioenergy feedstocks, there are many challenges in their direct utilization because of their recalcitrant chemistry. Lignocellulosic biomass primarily contains cellulose, hemicellulose and lignin in the ranges of 35–55 wt%, 20–40 wt% and 10–25 wt%, respectively (Nanda et al. 2013). A pretreatment step coupled with enzymatic hydrolysis and saccharification is necessary to denature the complex cellulose-hemicellulose-lignin framework, the matrix associated between lignin and various sugars so that microbial enzymes can access cellulose and hemicelluloses (Nanda et al. 2014c). Biomass pretreatment is achieved by a wide variety of mechanical (e.g. particle size reduction), physical (e.g. ultrasound, ozonolysis, microwave, irradiation, etc.), chemical (e.g. steam explosion, subcritical and supercritical fluids, acids, alkalis, liquid ammonia, ionic liquids, organosolv, etc.) and biological agents (e.g. cellulases, hemicellulases, lignin-modifying enzymes and lignin-degrading enzymes) (Nanda et al. 2014b).

During biomass pretreatment, the configuration of cellulosic fibers and highly branched arrangement of lignin is altered, thus facilitating the admittance of hydrolytic enzymes for saccharification and release of fermentable pentose and hexose sugars (Fougere et al. 2016; Rana et al. 2018b). There are many other benefits associated with pretreating lignocellulosic biomass for fermentative production of bioethanol and biobutanol, such as (i) faster hydrolysis, (ii) high product yields, (iii) reduced cellulose crystallinity, (iv) easier hemicellulose separation and (v) alteration and increase of biomass pore size for easy accessibility of cellulolytic enzymes (Nanda et al. 2015a).

4.3 ACETONE-BUTANOL-ETHANOL FERMENTATION

Acetone-Butanol-Ethanol (ABE) fermentation, mostly performed by *Clostridium* bacterium, has been explored for the conversion of several complex carbohydrates into biobutanol. *Saccharomyces cerevisiae*, which is a model fungus responsible for ethanol fermentation, lacks the natural ability to metabolize pentose sugars (i.e. hemicelluloses). Instead, it is efficient in metabolizing hexose sugars, mainly glucose (i.e. cellulose) through the glycolytic pathway (Walfridsson et al. 1995). On the contrary, *Clostridium* spp. can metabolize both pentose and hexose sugars, suggesting the utilization of hydrolyzed cellulose and hemicellulose from lignocellulosic biomass. This is another significant advantage of ABE fermentation over ethanol fermentation.

Clostridium is a rod-shaped gram-positive obligatory anaerobic bacterium that grows on a wide range of sugar substrates including starch, hemicellulose and cellulose. An array of enzymatic systems in *Clostridium* spp. enhances the production of biobutanol utilizing glucose, cellobiose, galactose, arabinose, mannose and xylose to butanol (Ezeji et al. 2007a). A few *Clostridium* spp. such as *C. acetobutylicum*, *C. beijerinckii*, *C. aurantibutyricum*, *C. butylicum*, *C. saccharobutylicum*, *C. saccharoperbutylacetonicum*, etc. have been explored for production of biobutanol through ABE fermentation (Dürre 2007; Nanda et al. 2017b). Some starch-based (first-generation) feedstocks like molasses, potatoes, corn, millet, rice, wheat and whey have the potential for butanol production, but their industrial usage is obsolete due to food *versus* fuel debate.

ABE fermentation is a biphasic bioconversion process consisting of acidogenic phase and solventogenic phase (Figure 4.1). In the acidogenic phase, the bacterium grows exponentially producing acetic acid and butyric acid from the sugars, whereas in the solventogenesis phase, the formation of acetone, butanol and ethanol takes place in a typical ratio of 3:6:1 (Dürre 2007). Some inhibition of the metabolic pathway occurs in the acetogenesis called as acidic stress, which favors the acid production rather than sugar consumption (Xue et al. 2013). The acidogenic phase is characterized by an increase in the acidity of the fermentation medium because of the formation of organic acids, which causes the bacterium to undergo the stationary growth phase and the subsequent solventogenic phase. The shift from the acetogenesis phase to solventogenic phase by *Clostridium* is characterized by a decelerated growth rate, formation of endospores and an increase in the levels of solvents (i.e. acetone, butanol and ethanol). For an obligate anaerobic bacterium, the acidogenic phase has a great role to play in its energy metabolism. As the pH level is reduced, the bacterium abridges acid formation

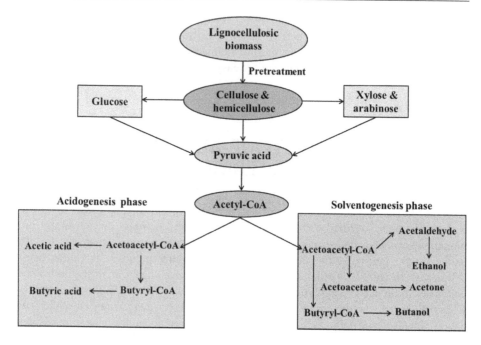

FIGURE 4.1 Simplified version of ABE fermentation by *Clostridium*

converting butyric acid and acetic acid to butanol and acetone, respectively. In the acidogenic phase, the redox equilibrium of butyric acid creates a balanced environment so that more butyric acid is formed than acetic acid (Zheng 2009).

According to the study conducted by García et al. (2011), shifting of phase during *Clostridium* metabolism occurs by the sporulation of 70–80% of the viable cells. A pH level of 5.5 facilitates the phase shifting from acidogenic phase to solventogenic phase. It should be noted that the decrease in the pH occurs in the late acidogenic phase by the accumulation of acetic acid and butyric acid (Lee et al. 2008). The role of pH in fermentation conditions is important for determining the production of acids and solvents. Nevertheless, by increasing the buffering capacity of the fermentation medium, the bacterial growth increases to support the utilization and conversion of remaining sugars, thereby producing more butanol (Nanda et al. 2017b).

The non-dissociation of butyric acid favors the phase-shifting (García et al. 2011) while the induction of butyrate kinase and acetate kinase occurs by butyric acid and acetic acid, respectively (Ballongue et al. 1986). Furthermore, butyryl-CoA along with butyryl phosphate facilitates the shifting of the acidogenic phase to the solventogenic phase. The ABE fermentation is accomplished by producing three types of products: (i) solvents such as acetone, butanol and ethanol, (ii) organic acids such as acetic acid, butyric acid and lactic acid and (iii) gases such as H_2 and CO_2 (Xue et al. 2013). The ABE fermentation period typically persists for 36–72 h producing approximately 20–25 g/L of total ABE (Qureshi and Ezeji 2008).

4.4 CHALLENGES AND OPPORTUNITIES OF ACETONE-BUTANOL-ETHANOL FERMENTATION

At different stages of ABE fermentation, some technical challenges are encountered. Although waste biomass is found to be an economical source of sugar substrates for fermentation, the main limitations are its total utilization by *Clostridium*, which is impacted due to butanol toxicity (inhibitory to the bacteria) and endospore formation (because of higher acid accumulation in the acidogenic phase). These limitations result in lower butanol yields by the bacteria (Ezeji et al. 2007b). The butanol yields and expensive solvent recovery methods (e.g. perstraction, pervaporation, gas stripping, liquid-liquid extraction, adsorption, distillation and supercritical fluid separation) make the ABE fermentation process expensive over bioethanol fermentation. The removal of inhibitors resulting from biomass pretreatment by detoxification is an important issue to address.

Butanol toxicity is a vital limitation of the industrial ABE fermentation process. *Clostridium* spp. seldom has a tolerance level of more than 2% butanol, which hampers the final yields of butanol through ABE fermentation. Butanol level up to 12–13 g/L is considered as the maximum limit for the wild-type *Clostridium* strains (García et al. 2011). For genetically modified *C. beijerinckii* BA101, the maximum yield of butanol reported is 19.6 g/L (Qureshi and Blaschek 1999). Apart from butanol concentration, the butanol recovery level is dependent on the bacteriophage infection. Bacteriophage such as *Siphoviridae* and *Podoviridae* have been reported to infect *C. madisonii* and *C. beijerinckii* P260, respectively (Jones et al. 2000).

The implementation of genetic and metabolic engineering is supported for overcoming some bottlenecks such as increased butanol tolerance by *Clostridium*. Another advancement like antisense RNA technology can be used for enhancing the microbial effectiveness for butanol production (Tummala et al. 2003). Therefore, synthetic biology approach can be applied for the development of novel microbial strains posing resistance to higher levels of butanol and contamination to bacteriophages, thereby sustaining the near-complete utilization of sugars and resulting in greater butanol yields.

4.5 CONCLUSIONS

Biobutanol has gained promising attention as an advanced alcohol-based biofuel. With many advantages such as the high-energy content and fuel properties similar to that of gasoline, biobutanol can be used as a next-generation biofuel in the automotive sectors. The longer hydrocarbon chains in biobutanol help for its use in existing internal combustion engines. As far as production of butanol production is concerned, ABE fermentation is a viable option producing acetone, butanol and ethanol in the ratio of 3:6:1. Apart from these three compounds, other byproducts obtained from ABE fermentation are acetic acid, butyric acid, CO_2 and H_2. Some major limitations of ABE fermentation have found to be butanol toxicity, incomplete sugar conversion and possibilities of bacteriophage infection.

C. acetobutylicum and *C. beijerinckii* have great potentials for biobutanol production through the ABE fermentation. Some technical challenges encountered during ABE fermentation are butanol toxicity, low butanol yields, difficulty in butanol separation due to its low yields, incomplete fermentation due to endospore formation and bacteriophage contamination. However, metabolic engineering of certain *Clostridium* sp. has proven to address some of these issues, thus leading to an improved butanol tolerance and yields. More research in bioprocess engineering is required to bring innovations to ABE fermentation and aid in an industrial scale sustainable production of butanol from a wide variety of lignocellulosic feedstocks and organic waste biomass.

REFERENCES

Azargohar, R., S. Nanda, K. Kang, T. Bond, C. Karunakaran, A.K. Dalai, J.A. Kozinski. 2019. Effects of bio-additives on the physicochemical properties and mechanical behavior of canola hull fuel pellets. *Renewable Energy* 132:296–307.

Ballongue, J., J. Amine, E. Masion, H. Petitdemange, D. Gay. 1986. Regulation of acetate kinase and butyrate kinase by acids in *Clostridium acetobutylicum*. *FEMS Microbiology Letters* 35:295–301.

Dürre, P. 2007. Biobutanol: an attractive biofuel. *Biotechnology Journal* 2:1525–1534.

Ezeji, T., N. Qureshi, H. Blaschek. 2007a. Butanol production from agricultural residues: impact of degradation products on *Clostridium beijerinckii* growth and butanol fermentation. *Biotechnology and Bioengineering* 97:1460–1469.

Ezeji, T.C., N. Qureshi, H.P. Blaschek. 2007b. Bioproduction of butanol from biomass: from genes to bioreactors. *Current Opinion in Biotechnology* 18:220–227.

Fougere, D., S. Nanda, K. Clarke, J.A. Kozinski, K. Li. 2016. Effect of acidic pretreatment on the chemistry and distribution of lignin in aspen wood and wheat straw substrates. *Biomass & Bioenergy* 91:56–68.

García, V., J. Päkkilä, H. Ojamo, E. Muurinen, R.L. Keiski. 2011. Challenges in biobutanol production: how to improve the efficiency? *Renewable and Sustainable Energy Reviews* 15:964–980.

Gong, M., S. Nanda, H.N. Hunter, W. Zhu, A.K. Dalai, J.A. Kozinski. 2017a. Lewis acid catalyzed gasification of humic acid in supercritical water. *Catalysis Today* 291:13–23.

Gong, M., S. Nanda, M.J. Romero, W. Zhu, J.A. Kozinski. 2017b. Subcritical and supercritical water gasification of humic acid as a model compound of humic substances in sewage sludge. *The Journal of Supercritical Fluids* 119:130–138.

Jones, D.T., M. Shirley, X. Wu, S. Keis. 2000. Bacteriophage infections in the industrial acetone butanol (AB) fermentation process. *Journal of Molecular Microbiology and Biotechnology* 2:21–26.

Kang, K., S. Nanda, S.S. Lam, T. Zhang, L. Huo, L. Zhao. 2020. Enhanced fuel characteristics and physical chemistry of microwave hydrochar for sustainable fuel pellet production via co-densification. *Environmental Research* 186:109480.

Lee, S.Y., J.H. Park, S.H. Jang, L.K. Nielsen, J. Kim, K.S. Jung. 2008. Fermentative butanol production by *Clostridia*. *Biotechnology Bioengineering* 101:209–228.

Nanda, S., A.K. Dalai, F. Berruti, J.A. Kozinski. 2016a. Biochar as an exceptional bioresource for energy, agronomy, carbon sequestration, activated carbon and specialty materials. *Waste and Biomass Valorization* 7:201–235.

Nanda, S., A.K. Dalai, I. Gökalp, J.A. Kozinski. 2016b. Valorization of horse manure through catalytic supercritical water gasification. *Waste Management* 52:147–158.

Nanda, S., A.K. Dalai, J.A. Kozinski. 2014a. Butanol and ethanol production from lignocellulosic feedstock: biomass pretreatment and bioconversion. *Energy Science & Engineering* 2:138–148.

Nanda, S., A.K. Dalai, J.A. Kozinski. 2016c. Supercritical water gasification of timothy grass as an energy crop in the presence of alkali carbonate and hydroxide catalysts. *Biomass & Bioenergy* 95:378–387.

Nanda, S., A.K. Dalai, J.A. Kozinski. 2017a. Butanol from renewable biomass: highlights on downstream processing and recovery techniques. In: *Sustainable Utilization of Natural Resources*; eds. P. Mondal, A.K. Dalai; 187–211. Boca Raton, FL: CRC Press.

Nanda, S., D. Golemi-Kotra, J.C. McDermott, A.K. Dalai, I. Gökalp, J.A. Kozinski. 2017b. Fermentative production of butanol: perspectives on synthetic biology. *New Biotechnology* 37:210–221.

Nanda, S., J. Isen, A.K. Dalai, J.A. Kozinski. 2016d. Gasification of fruit wastes and agro-food residues in supercritical water. *Energy Conversion and Management* 110:296–306.

Nanda, S., J. Maley, J.A. Kozinski, A.K. Dalai. 2015a. Physico-chemical evolution in lignocellulosic feedstocks during hydrothermal pretreatment and delignification. *Journal of Biobased Materials and Bioenergy* 9:295–308.

Nanda, S., J. Mohammad, S.N. Reddy, J.A. Kozinski, A.K. Dalai. 2014b. Pathways of lignocellulosic biomass conversion to renewable fuels. *Biomass Conversion and Biorefinery* 4:157–191.

Nanda, S., J.A. Kozinski, A.K. Dalai. 2016e. Lignocellulosic biomass: a review of conversion technologies and fuel products. *Current Biochemical Engineering* 3:24–36.

Nanda, S., M. Gong, H.N. Hunter, A.K. Dalai, I. Gökalp, J.A. Kozinski. 2017c. An assessment of pinecone gasification in subcritical, near-critical and supercritical water. *Fuel Processing Technology* 168:84–96.

Nanda, S., P. Mohanty, K.K. Pant, S. Naik, J.A. Kozinski, A.K. Dalai. 2013. Characterization of North American lignocellulosic biomass and biochars in terms of their candidacy for alternate renewable fuels. *Bioenergy Research* 6:663–677.

Nanda, S., R. Azargohar, A.K. Dalai, J.A. Kozinski. 2015b. An assessment on the sustainability of lignocellulosic biomass for biorefining. *Renewable and Sustainable Energy Reviews* 50:925–941.

Nanda, S., R. Azargohar, J.A. Kozinski, A.K. Dalai. 2014c. Characteristic studies on the pyrolysis products from hydrolyzed Canadian lignocellulosic feedstocks. *Bioenergy Research* 7:174–191.

Nanda, S., R. Rana, D.V.N. Vo, P.K. Sarangi, T.D. Nguyen, A.K. Dalai, J.A. Kozinski. 2020. A spotlight on butanol and propanol as next-generation synthetic fuels. In: *Biorefinery of Alternative Resources: Targeting Green Fuels and Platform Chemicals*; eds. S. Nanda, D.V.N. Vo, P.K. Sarangi; 105–126. Singapore: Springer Nature.

Nanda, S., R. Rana, H.N. Hunter, Z. Fang, A.K. Dalai, J.A. Kozinski. 2019a. Hydrothermal catalytic processing of waste cooking oil for hydrogen-rich syngas production. *Chemical Engineering Science* 195:935–945.

Nanda, S., R. Rana, P.K. Sarangi, A.K. Dalai, J.A. Kozinski. 2018a. A broad introduction to first, second and third generation biofuels. In: *Recent Advancements in Biofuels and Bioenergy Utilization*; eds. P.K. Sarangi, S. Nanda, P. Mohanty; 1–25. Singapore: Springer Nature.

Nanda, S., R. Rana, Y. Zheng, J.A. Kozinski, A.K. Dalai. 2017d. Insights on pathways for hydrogen generation from ethanol. *Sustainable Energy & Fuels* 1:1232–1245.

Nanda, S., S.N. Reddy, A.K. Dalai, J.A. Kozinski. 2016f. Subcritical and supercritical water gasification of lignocellulosic biomass impregnated with nickel nanocatalyst for hydrogen production. *International Journal of Hydrogen Energy* 41:4907–4921.

Nanda, S., S.N. Reddy, D.V.N. Vo, B.N. Sahoo, J.A. Kozinski. 2018b. Catalytic gasification of wheat straw in hot compressed (subcritical and supercritical) water for hydrogen production. *Energy Science & Engineering* 6:448–459.

Nanda, S., S.N. Reddy, H.N. Hunter, A.K. Dalai, J.A. Kozinski. 2015c. Supercritical water gasification of fructose as a model compound for waste fruits and vegetables. *The Journal of Supercritical Fluids* 104:112–121.

Nanda, S., S.N. Reddy, H.N. Hunter, D.V.N. Vo, J.A. Kozinski, I. Gökalp. 2019b. Catalytic subcritical and supercritical water gasification as a resource recovery approach from waste tires for hydrogen-rich syngas production. *The Journal of Supercritical Fluids* 154:104627.

Nanda, S., S.N. Reddy, H.N. Hunter, I.S. Butler, J.A. Kozinski. 2015d. Supercritical water gasification of lactose as a model compound for valorization of dairy industry effluents. *Industrial & Engineering Chemistry Research* 54:9296–9306.

Nanda, S., S.N. Reddy, S.K. Mitra, J.A. Kozinski. 2016g. The progressive routes for carbon capture and sequestration. *Energy Science & Engineering* 4:99–122.

Okolie, J.A., S. Nanda, A.K. Dalai, F. Berruti, J.A. Kozinski. 2020a. A review on subcritical and supercritical water gasification of biogenic, polymeric and petroleum wastes to hydrogen-rich synthesis gas. *Renewable and Sustainable Energy Reviews* 119:109546.

Okolie, J.A., S. Nanda, A.K. Dalai, J.A. Kozinski. 2020b. Chemistry and specialty industrial applications of lignocellulosic biomass. *Waste and Biomass Valorization.* Doi: 10.1007/s12649-020-01123-0.

Okolie, J.A., S. Nanda, A.K. Dalai, J.A. Kozinski. 2020c. Hydrothermal gasification of soybean straw and flax straw for hydrogen-rich syngas production: experimental and thermodynamic modeling. *Energy Conversion and Management* 208:112545.

Parakh, P.D., S. Nanda, J.A. Kozinski. 2020. Eco-friendly transformation of waste biomass to biofuels. *Current Biochemical Engineering* 6:120–134.

Qureshi, N., H.P. Blaschek. 1999. Production of acetone butanol ethanol (ABE) by a hyper-producing mutant strain of *Clostridium beijerinckii* BA101 and recovery by pervaporation. *Biotechnology Progress* 15:594–602.

Qureshi, N., T.C. Ezeji. 2008. Butanol, 'a superior biofuel' production from agricultural residues (renewable biomass): recent progress in technology. *Biofuels, Bioproducts and Biorefinery* 2:319–330.

Rana, R., S. Nanda, A. Maclennan, Y. Hu, J.A. Kozinski, A.K. Dalai. 2019. Comparative evaluation for catalytic gasification of petroleum coke and asphaltene in subcritical and supercritical water. *Journal of Energy Chemistry* 31:107–118.

Rana, R., S. Nanda, J.A. Kozinski, A.K. Dalai. 2018a. Investigating the applicability of Athabasca bitumen as a feedstock for hydrogen production through catalytic supercritical water gasification. *Journal of Environmental Chemical Engineering* 6:182–189.

Rana, R., S. Nanda, S.N. Reddy, A.K. Dalai, J.A. Kozinski, I. Gökalp. 2020. Catalytic gasification of light and heavy gas oils in supercritical water. *Journal of the Energy Institute* 93:2025-2032.

Rana, R., S. Nanda, V. Meda, A.K. Dalai, J.A. Kozinski. 2018b. A review of lignin chemistry and its biorefining conversion technologies. *Journal of Biochemical Engineering and Bioprocess Technology* 1:2.

Sarangi, P.K., S. Nanda. 2018. Recent developments and challenges of acetone-butanol-ethanol fermentation. In: *Recent Advancements in Biofuels and Bioenergy Utilization*; eds. P.K. Sarangi, S. Nanda, P. Mohanty; 111–123. Singapore: Springer Nature.

Sarangi, P.K., S. Nanda. 2019a. Bioconversion of agro-wastes into phenolic flavour compounds. In: *Biotechnology for Sustainable Energy and Products*; eds. P.K. Sarangi, S. Nanda; 266–284. New Delhi, India: I.K. International Publishing House Pvt. Ltd.

Sarangi, P.K., S. Nanda. 2019b. Recent advances in consolidated bioprocessing for microbe-assisted biofuel production. In: *Fuel Processing and Energy Utilization*; eds. S. Nanda, P.K. Sarangi, D.V.N. Vo; 141–157. Boca Raton, FL: CRC Press.

Sarangi, P.K., S. Nanda. 2020. Biohydrogen production through dark fermentation. *Chemical Engineering & Technology* 43:601–612.

Sarangi, P.K., S. Nanda, D.V.N. Vo. 2020. Technological advancements in the production and application of biomethanol. In: *Biorefinery of Alternative Resources: Targeting Green Fuels and Platform Chemicals*; eds. S. Nanda, D.V.N. Vo, P.K. Sarangi; 127–139. Singapore: Springer Nature.

Singh, A., S. Nanda, F. Berruti. 2020. A review of thermochemical and biochemical conversion of miscanthus to biofuels. In: *Biorefinery of Alternative Resources: Targeting Green Fuels and Platform Chemicals*; eds. S. Nanda, D.V.N. Vo, P.K. Sarangi; 195–220. Singapore: Springer Nature.

Sun, J., L. Xu, G.H. Dong, S. Nanda, H. Li, Z. Fang, J.A. Kozinski, A.K. Dalai. 2020. Subcritical water gasification of lignocellulosic wastes for hydrogen production with Co modified Ni/Al$_2$O$_3$ catalyst. *The Journal of Supercritical Fluids* 162:104863.

Tummala, S.B., N.E. Welker, E.T. Papoutsakis. 2003. Design of antisense RNA constructs for downregulation of the acetone formation pathway of *Clostridium acetobutylicum*. *Journal of Bacteriology* 185:1923–1934.

Walfridsson, M., J. Hallborn, M. Penttilä, S. Keränen, B. Hahn-Hägerdal. 1995. Xylose-metabolizing *Saccharomyces cerevisiae* strains overexpressing the TKL1 and TAL1 genes encoding the pentose phosphate pathway enzymes transketolase and transaldolase. *Applied and Environmental Microbiology* 61:4184–4190.

Xue, C., X.Q. Zhao, C.G. Liu, J. Chen, F.W. Bai. 2013. Prospective and development of butanol as an advanced biofuel. *Biotechnology Advances* 31:1575–1584.

Zheng, Y.N., L.Z. Li, M. Xian, Y.J. Ma, J.M. Yang, X. Xu, D.Z. He. 2009. Problems with the microbial production of butanol. *Journal of Industrial Microbiology and Biotechnology* 36:1127–1138.

Bioconversion of Waste Biomass to Biomethanol

5

5.1 INTRODUCTION

Today, the major environmental concerns are focused on the effective production and application of biofuels, biochemicals, biomaterials and bioproducts to replace fossil fuels and its derivatives at a global scale (Nanda et al. 2016d; Parakh et al. 2020; Okolie et al. 2020b). Because of massive amounts of greenhouse gas emissions, environmental pollution, global warming and climate change, rising crude oil prices and high carbon taxes associated with the exploiting usage of fossil fuels, there is a global impetus to seek eco-friendly and alternative renewable resources (Nanda et al. 2015; Rana et al. 2018; Rana et al. 2019; Rana et al. 2020; Okolie et al. 2020a). Harnessing the hydrocarbons present in the waste lignocellulosic biomass as carbohydrates is of great potential to sustain their conversion to biofuels and biochemicals (Nanda et al. 2013; Sarangi et al. 2017; Azargohar et al. 2018; Sarangi et al. 2018; Azargohar et al. 2019; Kang et al. 2019; Sarangi and Nanda 2019a; Sarangi and Nanda 2019b; Kang et al. 2020).

Lignocellulosic biomass containing cellulose, hemicellulose and lignin can be transformed to biofuels using several thermochemical technologies (e.g. pyrolysis, liquefaction, gasification, transesterification, torrefaction, reforming, etc.) and biological technologies (e.g. anaerobic digestion, enzymatic saccharification, photo-fermentation, dark fermentation, acetone-butanol-ethanol fermentation, ethanol fermentation, syngas fermentation, etc.) (Nanda et al. 2014b; Nanda et al. 2017a; Nanda et al. 2017b; Nanda et al. 2017c; Nanda et al. 2017e; Nanda et al. 2018; Sarangi and Nanda 2018; Sarangi and Nanda 2020). Based on the conversion methods, various fuel products obtained from lignocellulosic biomasses are bio-oil, biodiesel, biomethanol, bioethanol, biobutanol, biopropanol, biohydrogen, biomethane, syngas, etc. (Reddy et al. 2014; Nanda et al. 2014a; Reddy et al. 2016; Nanda et al. 2016c; Reddy et al. 2017; Reddy et al. 2018; Nayak et al. 2019; Okolie et al. 2019; Reddy et al. 2019; Yadav et al. 2019; Nanda et al. 2020; Nayak et al. 2020; Sarangi et al. 2020). Bioethanol, biobutanol, biomethanol and biopropanol are some of the alcohol-based bioproducts obtained predominantly from the microbial bioprocessing of waste biomass with the potential to be used as biofuels or commodity biochemicals.

Biomethanol, being one of the most dynamic and vibrant fuel substitutes, can be generated from waste biomass by utilizing specific microbial communities (Sarangi et al. 2020). With diverse applications in several industrial sectors worldwide, the environmental-friendly production of biomethanol is gaining global interest in research, development and innovation. This chapter describes some notable value-added applications of biomethanol as well as its fermentative production from waste biomass using selective microorganisms.

5.2 APPLICATIONS OF BIOMETHANOL

Biomethanol (CH_3OH) has many recognized applications in chemicals, fuels and other specialty sectors. Biomethanol is widely applied in the following sectors such as: (i) transportation fuels, (ii) blending with gasoline and diesel, (iii) conversion into dimethyl ether for diesel alternatives, (iv) electricity from fuel cells and (v) biodiesel production through transesterification process (Yanju et al. 2008; Matzen and Demirel 2016; Reddy et al. 2016; Reddy et al. 2018; Bhatia et al. 2020; Sarangi et al. 2020). In all the above-mentioned applications, a key attribute of biomethanol is associated with its sustainability, carbon-neutrality (if produced from bioresources) and cost-effectiveness compared to other alcohol-based hydrocarbon fuels.

Fuel properties such as research octane number, motor octane number and anti-knock index of methanol are nearly 109, 89 and 99, respectively (Eyidogan et al. 2010). On the other hand, the research octane number and motor octane number of gasoline are in the range of 91–99 and 81–89, respectively (Nanda et al. 2017b). Having a greater octane number than gasoline, methanol is considered as a promising low-cost alternative fuel with satisfactory fuel performance and less environmental impacts (Fatih et al. 2011).

The blending of biomethanol is one of its significant applications found in the automobile sectors. Besides, blending 15% methanol with gasoline and 20% methanol with diesel requires slight modifications to the vehicular engines for use as a transportation fuel (Kowalewicz 1993). Moreover, methanol-ethanol-gasoline blended fuels can enhance engine performance and greater efficiencies along with lower CO and NO_x emissions than that of gasoline alone (Elfasakhany 2015).

Methanol production from CO_2 is a sustainable option for recycling CO_2 and reducing its concentration as a potent greenhouse gas from the atmosphere (Nguyen et al. 2020). Methanol can be produced using CO_2 from industrial flue gas. Nearly about 0.19 t of methanol was produced per ton of fossil fuel, thereby resulting in a reduction of 0.42 MT of CO_2 emissions per year (Ptasinski et al. 2002). The application of biomethanol is also found in the power generation sector for gas turbines (Galindo and Badr 2007; Suntana et al. 2009). In another attribute, methanol can be used for biodiesel production via transesterification (Reddy et al. 2018). Besides, residual methanol and glycerol obtained as waste effluents from biodiesel industries can serve as a precursor for hydrogen production through supercritical water gasification (Reddy et al. 2016). Methanol is also used as an anti-frost agent, organic solvent and precursor for producing several fine chemicals (Sarangi et al. 2020).

5.3 PRODUCTION OF BIOMETHANOL

The conversion of biomass to methanol is achieved through catalytic thermochemical processes and biological processes mediated by methanotrophic bacteria. In the thermochemical processes, waste biomass undergoes gasification to produce synthesis gas or syngas, which comprises CO, CO_2, H_2, CH_4 and traces of C_{2+} gases (Nanda et al. 2016a; Nanda et al. 2016b; Nanda et al. 2017d; Okolie et al. 2020c). In the next step, the conditioning of syngas is performed to remove various impurities such as tar and other undesired gases to optimize the ratio of H_2:CO (Okolie et al. 2019). Through Fischer-Tropsch catalysis, the conditioned syngas is converted to hydrocarbon fuels, chemicals and alcohols including methanol (Venvik and Yang 2017; Singh et al. 2018). The conversion of pectin produces uronic acid that aids in methanol production by combining with ether (Bhattacharyya et al. 2008). Another novel approach for the production of biomethanol from bio-oil has been reported through CO-rich bio-syngas, which in turn was produced from CO_2-rich bio-syngas (Xu et al. 2011).

On the other hand, the biological production of methanol deals with the utilization of waste biomass and anaerobic bioprocesses mediated by methanotrophic bacteria. Lignocellulosic biomass including agricultural crop residues and forestry refuses to act as promising resources and storehouses of fermentable hexose and pentose sugars for conversion to biofuels and biochemicals. Methanotrophic bacteria have been explored for the conversion of methane to methanol (Hanson and Hanson 1996). A few examples of methanotrophic bacteria are *Methylocaldum*, *Methylococcus*, *Methylogaea*, *Methylohalobius*, *Methylomarinovum*, *Methyloparacoccus*, *Methylothermus*, etc. (Bjorck et al. 2018). *Bacillus methanicus* is one of the notable methanotrophic bacteria.

Depending on the availability of methane in the environment, two population types of methanotrophs have been considered for methane conversion to methanol (Bender and Conrad 1992). The first category of methanotrophs is found in soils having high methane concentration. Such microorganisms convert methane at a level of more than 40 ppm concentration, thereby being regarded as the low-affinity methanotrophs. On the other hand, the second category of methanotrophs grows at a low methane concentration of about 2 ppm, thereby being known as the high-affinity methanotrophs. By the catalytic action of methane monooxygenase (MMO), the methanotrophs convert methane into methanol through the oxidation reaction (Figure 5.1).

Soluble cytoplasmic form (sMMO) as well as particulate membrane-bound form (pMMO) are the two forms of methane monooxygenase. There are other possible routes for the conversion of methanol into CO_2 via formaldehyde and formic acid

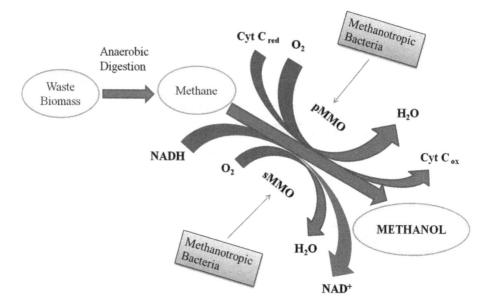

FIGURE 5.1   Conversion of methane to methanol via methanotrophic bacteria

and three different types of enzymes, i.e. methanol dehydrogenase (MDH), form-aldehyde dehydrogenase (FADH) and formate dehydrogenase (FDH) (Hanson and Hanson 1996; Xin et al. 2009). The selection of the methanotrophic bacteria and the standardization of its growth conditions can affect the final recovery of biomethanol. More exploration of potential microbial communities and understanding their bio-catalytic activities could aid in the large-scale production of biomethanol from waste biomass sources.

5.4 CONCLUSIONS

The implementation of microbial communities to utilize waste biomass sources for the production of biofuels and biochemicals is a sustainable alternative to extraction, pro-cessing and utilization of fossil fuels. Waste biomass can act as an alternative feedstock for the production of biomethanol by methanotrophic bacteria via methane conversion. Various applications of methanol have been recognized in fuels, chemicals and other specialty applications. Noteworthy developments in the biological conversion processes could result in large-scale commercial production of biomethanol for industrial applica-tions. Moreover, methane is considered as the second most potent greenhouse gas after CO_2. Hence, its conversion to biomethanol has great advantages in reducing its atmo-spheric concentrations and mitigating global warming.

REFERENCES

Azargohar, R., S. Nanda, A.K. Dalai. 2018. Densification of agricultural wastes and forest residues: a review on influential parameters and treatments. In: *Recent Advancements in Biofuels and Bioenergy Utilization*; eds. P.K. Sarangi, S. Nanda, P. Mohanty; 27–51. Singapore: Springer Nature.

Azargohar, R., S. Nanda, K. Kang, T. Bond, C. Karunakaran, A.K. Dalai, J.A. Kozinski. 2019. Effects of bio-additives on the physicochemical properties and mechanical behavior of canola hull fuel pellets. *Renewable Energy* 132:296–307.

Bender, M., R. Conrad. 1992. Kinetics of CH_4 oxidation in oxic soils exposed to ambient air or high CH_4 mixing ratios. *FEMS Microbiology Ecology* 101:261–270.

Bhatia, L., P.K. Sarangi, S. Nanda. 2020. Current advancements in microbial fuel cell technologies. In: *Biorefinery of Alternative Resources: Targeting Green Fuels and Platform Chemicals*; eds. S. Nanda, D.V.N. Vo, P.K. Sarangi; 477–494. Singapore: Springer Nature.

Bhattacharyya, J.K., S. Kumar, S. Devotta. 2008. Studies on acidification in two-phase biomethanation process of municipal solid waste. *Waste Management* 28:164–169.

Bjorck, C.E., P.D. Dobson, J. Pandhal. 2018. Biotechnological conversion of methane to methanol: evaluation of progress and potential. *AIMS Bioengineering* 5:1–38.

Elfasakhany, A. 2015. Investigations on the effects of ethanol–methanol–gasoline blends in a spark-ignition engine: performance and emissions analysis. *Engineering Science and Technology, an International Journal* 18:713–719.

Eyidogan, M., A.N. Ozsezen, M. Canakci, A. Turkcan. 2010. Impact of alcohol–gasoline fuel blends on the performance and combustion characteristics of an SI engine. *Fuel* 89:2713–2720.

Fatih, D.M., M. Balat, H. Balat. 2011. Biowastes to biofuels. *Energy Conversion and Management* 52:1815–1828.

Galindo, C.P., O. Badr. 2007. Renewable hydrogen utilisation for the production of methanol. *Energy Conversion and Management* 48:519–527.

Hanson, R.S., T.E. Hanson. 1996. Methanotrophic bacteria. *Microbial Reviews* 60:439–471.

Kang, K., S. Nanda, G. Sun, L. Qiu, Y. Gu, T. Zhang, M. Zhu, R. Sun. 2019. Microwave-assisted hydrothermal carbonization of corn stalk for solid biofuel production: optimization of process parameters and characterization of hydrochar. *Energy* 186:115795.

Kang, K., S. Nanda, S.S. Lam, T. Zhang, L. Huo, L. Zhao. 2020. Enhanced fuel characteristics and physical chemistry of microwave hydrochar for sustainable fuel pellet production via co-densification. *Environmental Research* 186:109480.

Kowalewicz, A. 1993. Methanol as a fuel for spark ignition engines: a review and analysis. *Proceedings of the Institution of Mechanical Engineers, Part D: Journal of Automobile Engineering* 207:43–52.

Matzen, M., Y. Demirel. 2016. Methanol and dimethyl ether from renewable hydrogen and carbon dioxide: alternative fuels production and life-cycle assessment. *Journal of Cleaner Production* 139:1068–1077.

Nanda, S., A.K. Dalai, I. Gökalp, J.A. Kozinski. 2016a. Valorization of horse manure through catalytic supercritical water gasification. *Waste Management* 52:147–158.

Nanda, S., A.K. Dalai, J.A. Kozinski. 2014a. Butanol and ethanol production from lignocellulosic feedstock: biomass pretreatment and bioconversion. *Energy Science & Engineering* 2:138–148.

Nanda, S., A.K. Dalai, J.A. Kozinski. 2016b. Supercritical water gasification of timothy grass as an energy crop in the presence of alkali carbonate and hydroxide catalysts. *Biomass & Bioenergy* 95:378–387.

Nanda, S., A.K. Dalai, J.A. Kozinski. 2017a. Butanol from renewable biomass: highlights on downstream processing and recovery techniques. In: *Sustainable Utilization of Natural Resources*; eds. P. Mondal, A.K. Dalai; 187–211. Boca Raton, FL: CRC Press.

Nanda, S., D. Golemi-Kotra, J.C. McDermott, A.K. Dalai, I. Gökalp, J.A. Kozinski. 2017b. Fermentative production of butanol: perspectives on synthetic biology. *New Biotechnology* 37:210–221.

Nanda, S., J. Mohammad, S.N. Reddy, J.A. Kozinski, A.K. Dalai. 2014b. Pathways of lignocellulosic biomass conversion to renewable fuels. *Biomass Conversion and Biorefinery* 4:157–191.

Nanda, S., J.A. Kozinski, A.K. Dalai. 2016c. Lignocellulosic biomass: a review of conversion technologies and fuel products. *Current Biochemical Engineering* 3:24–36.

Nanda, S., K. Li, N. Abatzoglou, A.K. Dalai, J.A. Kozinski. 2017c. Advancements and confinements in hydrogen production technologies. In: *Bioenergy Systems for the Future*; eds. F. Dalena, A. Basile, C. Rossi; 373–418. Cambridge: Woodhead Publishing: Elsevier.

Nanda, S., M. Gong, H.N. Hunter, A.K. Dalai, I. Gökalp, J.A. Kozinski. 2017d. An assessment of pinecone gasification in subcritical, near-critical and supercritical water. *Fuel Processing Technology* 168:84–96.

Nanda, S., P. Mohanty, K.K. Pant, S. Naik, J.A. Kozinski, A.K. Dalai. 2013. Characterization of North American lignocellulosic biomass and biochars in terms of their candidacy for alternate renewable fuels. *Bioenergy Research* 6:663–677.

Nanda, S., R. Azargohar, A.K. Dalai, J.A. Kozinski. 2015. An assessment on the sustainability of lignocellulosic biomass for biorefining. *Renewable and Sustainable Energy Reviews* 50:925–941.

Nanda, S., R. Rana, D.V.N. Vo, P.K. Sarangi, T.D. Nguyen, A.K. Dalai, J.A. Kozinski. 2020. A spotlight on butanol and propanol as next-generation synthetic fuels. In: *Biorefinery of Alternative Resources: Targeting Green Fuels and Platform Chemicals*; eds. S. Nanda, D.V.N. Vo, P.K. Sarangi; 105–126. Singapore: Springer Nature.

Nanda, S., R. Rana, P.K. Sarangi, A.K. Dalai, J.A. Kozinski. 2018. A broad introduction to first, second and third generation biofuels. In: *Recent Advancements in Biofuels and Bioenergy Utilization*; eds. P.K. Sarangi, S. Nanda, P. Mohanty; 1–25. Singapore: Springer Nature.

Nanda, S., R. Rana, Y. Zheng, J.A. Kozinski, A.K. Dalai. 2017e. Insights on pathways for hydrogen generation from ethanol. *Sustainable Energy & Fuels* 1:1232–1245.

Nanda, S., S.N. Reddy, S.K. Mitra, J.A. Kozinski. 2016d. The progressive routes for carbon capture and sequestration. *Energy Science & Engineering* 4:99–122.

Nayak, S.K., B. Nayak, P.C. Mishra, M.M. Noor, S. Nanda. 2019. Effects of biodiesel blends and producer gas flow on overall performance of a turbocharged direct injection dual-fuel engine. *Energy Sources, Part A: Recovery, Utilization, and Environmental Effects.* Doi: 10.1080/15567036.2019.1694101.

Nayak, S.K., P.C. Mishra, S. Nanda, B. Nayak, M.M. Noor. 2020. Opportunities for biodiesel compatibility as a modern combustion engine fuel. In: *Biorefinery of Alternative Resources: Targeting Green Fuels and Platform Chemicals*; eds. S. Nanda, D.V.N. Vo, P.K. Sarangi; 457–476. Singapore: Springer Nature.

Nguyen, T.D., T.V. Tran, S. Singh, P.T.T. Phuong, L.G. Bach, S. Nanda, D.V.N. Vo. 2020. Conversion of carbon dioxide into formaldehyde. In: *Conversion of Carbon Dioxide into Hydrocarbons. Vol. 2 Technology*; eds. A.M. Asiri, Inamuddin, L. Eric; 159–183. Singapore: Springer Nature.

Okolie, J.A., R. Rana, S. Nanda, A.K. Dalai, J.A. Kozinski. 2019. Supercritical water gasification of biomass: a state-of-the-art review of process parameters, reaction mechanisms and catalysis. *Sustainable Energy & Fuels* 3:578–598.

Okolie, J.A., S. Nanda, A.K. Dalai, F. Berruti, J.A. Kozinski. 2020a. A review on subcritical and supercritical water gasification of biogenic, polymeric and petroleum wastes to hydrogen-rich synthesis gas. *Renewable and Sustainable Energy Reviews* 119:109546.

Okolie, J.A., S. Nanda, A.K. Dalai, J.A. Kozinski. 2020b. Chemistry and specialty industrial applications of lignocellulosic biomass. *Waste and Biomass Valorization*. Doi: 10.1007/s12649-020-01123-0.

Okolie, J.A., S. Nanda, A.K. Dalai, J.A. Kozinski. 2020c. Hydrothermal gasification of soybean straw and flax straw for hydrogen-rich syngas production: experimental and thermodynamic modeling. *Energy Conversion and Management* 208:112545.

Parakh, P.D., S. Nanda, J.A. Kozinski. 2020. Eco-friendly transformation of waste biomass to biofuels. *Current Biochemical Engineering* 6:120–134.

Ptasinski, K.J., C. Hamelinck, P.J.A.M. Kerkhof. 2002. Exergy analysis of methanol from the sewage sludge process. *Energy Conversion and Management* 43:1445–1457.

Rana, R., S. Nanda, A. Maclennan, Y. Hu, J.A. Kozinski, A.K. Dalai. 2019. Comparative evaluation for catalytic gasification of petroleum coke and asphaltene in subcritical and supercritical water. *Journal of Energy Chemistry* 31:107–118.

Rana, R., S. Nanda, J.A. Kozinski, A.K. Dalai. 2018. Investigating the applicability of Athabasca bitumen as a feedstock for hydrogen production through catalytic supercritical water gasification. *Journal of Environmental Chemical Engineering* 6:182–189.

Rana, R., S. Nanda, S.N. Reddy, A.K. Dalai, J.A. Kozinski, I. Gökalp. 2020. Catalytic gasification of light and heavy gas oils in supercritical water. *Journal of the Energy Institute* 93:2025-2032.

Reddy, S.N., S. Nanda, A.K. Dalai, J.A. Kozinski. 2014. Supercritical water gasification of biomass for hydrogen production. *International Journal of Hydrogen Energy* 39:6912–6926.

Reddy, S.N., S. Nanda, J.A. Kozinski. 2016. Supercritical water gasification of glycerol and methanol mixtures as model waste residues from biodiesel refinery. *Chemical Engineering Research and Design* 113:17–27.

Reddy, S.N., S. Nanda, P. Kumar, M.C. Hicks, U.G. Hegde, J.A. Kozinski. 2019. Impacts of oxidant characteristics on the ignition of n-propanol-air hydrothermal flames in supercritical water. *Combustion and Flame* 203:46–55.

Reddy, S.N., S. Nanda, P.K. Sarangi. 2018. Applications of supercritical fluids for biodiesel production. In: *Recent Advancements in Biofuels and Bioenergy Utilization*; eds. P.K. Sarangi, S. Nanda, P. Mohanty; 261–284. Singapore: Springer Nature.

Reddy, S.N., S. Nanda, U.G. Hegde, M.C. Hicks, J.A. Kozinski. 2017. Ignition of n-propanol–air hydrothermal flames during supercritical water oxidation. *Proceedings of the Combustion Institute* 36:2503–2511.

Sarangi, P.K., S. Nanda. 2018. Recent developments and challenges of acetone-butanol-ethanol fermentation. In: *Recent Advancements in Biofuels and Bioenergy Utilization*; eds. P.K. Sarangi, S. Nanda, P. Mohanty; 111–123. Singapore: Springer Nature.

Sarangi, P.K., S. Nanda. 2019a. Bioconversion of agro-wastes into phenolic flavour compounds. In: *Biotechnology for Sustainable Energy and Products*; eds. P.K. Sarangi, S. Nanda; 266–284. New Delhi: I.K. International Publishing House Pvt. Ltd.

Sarangi, P.K., S. Nanda. 2019b. Recent advances in consolidated bioprocessing for microbe-assisted biofuel production. In: *Fuel Processing and Energy Utilization*; eds. S. Nanda, P.K. Sarangi, D.V.N. Vo; 141–157. Boca Raton, FL: CRC Press.

Sarangi, P.K., S. Nanda. 2020. Biohydrogen production through dark fermentation. *Chemical Engineering & Technology* 43:601–612.

Sarangi, P.K., S. Nanda, D.V.N. Vo. 2020. Technological advancements in the production and application of biomethanol. In: *Biorefinery of Alternative Resources: Targeting Green Fuels and Platform Chemicals*; eds. S. Nanda, D.V.N. Vo, P.K. Sarangi; 127–139. Singapore: Springer Nature.

Sarangi, P.K., S. Nanda, R. Das. 2018. Bioconversion pineapple residues for recovery of value-added compounds. *International Journal of Research in Science and Engineering* 210–217.

Sarangi, P.K., T.A. Singh, J. Singh. 2017. Agricultural crop residues: unutilized biomass having huge energy potential. In: *Contemporary Renewable Energy Technologies for Sustainable Agriculture*; ed. M. Ghosal; 47–61. New Delhi: Narosa Publishing House.

Singh, S., R. Kumar, H.D. Setiabudi, S. Nanda, D.V.N. Vo. 2018. Advanced synthesis strategies of mesoporous SBA-15 supported catalysts for catalytic reforming applications: a state-of-the-art review. *Applied Catalysis A: General* 559:57–74.

Suntana, A.S., K.A. Vogt, E.C. Turnblom, U. Ravi. 2009. Bio-methanol potential in Indonesia: forest biomass as a source of bio-energy that reduces carbon emissions. *Applied Energy* 86:215–221.

Venvik, H.J., J. Yang. 2017. Catalysis in microstructured reactors: short review on small-scale syngas production and further conversion into methanol, DME and Fischer-Tropsch products. *Catalysis Today* 285:135–146.

Xin, J.Y., J.R. Cui, J.Z. Niu, S.F. Hua, Z.G. Xia, S.B. Li, L.M. Zhu. 2009. Production of methanol from methane by methanotrophic bacteria. *Biocatalysis and Biotransformation* 22:225–229.

Xu, Y., T. Ye, S. Qiu, S. Ning, F. Gong, Y. Liu, Q. Li. 2011. High efficient conversion of CO_2-rich bio-syngas to CO-rich bio-syngas using biomass char: a useful approach for production of bio-methanol from bio-oil. *Bioresource Technology* 102:6239–6245.

Yadav, P., S.N. Reddy, S. Nanda. 2019. Cultivation and conversion of algae for wastewater treatment and biofuel production. In: *Fuel Processing and Energy Utilization*; eds. S. Nanda, P.K. Sarangi, D.V.N. Vo; 159–175. Boca Raton, FL: CRC Press.

Yanju, W., L. Shenghua, L. Hongsong, Y. Rui, L. Jie, W. Ying. 2008. Effects of methanol/gasoline blends on a spark ignition engine performance and emissions. *Energy & Fuels* 22:1254–1259.

Bioconversion of Waste Biomass to Biohydrogen

6

6.1 INTRODUCTION

The energy carriers and chemicals are exclusively dependent on fossil fuel sources in the present world scenario. The use of fossil fuels not only increases the concentration of greenhouse gases in the atmosphere but also leads to environmental degradation, pollution, global warming, climate change, acid rain, extreme weather patterns, volatility in the crude oil prices and occasional geopolitical tensions between fuel importing and exporting nations (Nanda et al. 2016b; Rana et al. 2018; Rana et al. 2019; Rana et al. 2020; Vakulchuk et al. 2020). With many dependencies on fossil fuel energy sources such as crude oil, gasoline, diesel, coal and natural gas, the major global concern is to seek an alternative energy carrier and vector that can make a paradigm shift in the energy usage from fossil fuels to biofuels for long-term scenarios (Nanda et al. 2015b). Among all biofuel sources, hydrogen has great potentials for use as an outstanding energy carrier and vector. Being a next-generation biofuel, hydrogen can be utilized as a direct fuel, in fuel cells or as a precursor to producing advanced hydrocarbon fuels and chemicals (Nanda et al. 2017b; Okolie et al. 2019; Bhatia et al. 2020; Sarangi and Nanda 2020; Okolie et al. 2020a).

There is an increasing interest in hydrogen generation and utilization as a transportation fuel, as a mitigation strategy to the harmful impact of fossil fuels on the environment. Hydrogen (H_2) is often referred to as the "fuel of the future" (Reddy et al. 2020). Having a thermal efficiency of about 120 MJ/kg and flame temperature of 2027°C, hydrogen shows an energy content of several magnitudes higher than that of other conventional hydrocarbon fuels (Nanda et al. 2017c). Hydrogen is an environmental-friendly fuel having versatile applications in energy, fuel, combined heat and power, chemical, metallurgy, fertilizers and other commercial and industrial sectors. Hydrogen is considered as a promising renewable energy source for the sustainable future because its combustion generates massive amounts of heat energy and water.

Hydrogen can be produced using a variety of pathways involving thermochemical technologies (e.g. gasification, pyrolysis and reforming), electrolysis, electrochemical, photochemical, photo-catalytic, photo-electrochemical and microbial (photolysis, photo-fermentation, dark fermentation and microbial electrolysis cells) (Nanda et al. 2017b; Huang et al. 2020; Siang et al. 2020). This chapter describes the microbial technologies for biohydrogen production.

6.2 BIOLOGICAL METHODS FOR HYDROGEN PRODUCTION

The production of biohydrogen is considered beneficial as far as ecological and energy requirements are concerned. The biological method for biohydrogen production demonstrates less negative impacts on the environment and low requirement of energy as compared to other sources (Nanda et al. 2017b; Sarangi and Nanda 2020). Lignocellulosic biomass contains major fractions of celluloses and hemicelluloses, which could act as the sources of pentose and hexose sugars for microbial conversion to fuels and chemicals (Nanda et al. 2013; Nanda et al. 2014a; Nanda et al. 2017a; Sarangi and Nanda 2018; Sarangi and Nanda 2019; Sarangi et al. 2020). Before the fermentation process, biomass is required to be pretreated and delignified, to remove the lignin and release the monomeric sugars for fermentation to fuels and value-added chemicals (Nanda et al. 2014b; Nanda et al. 2015a; Fougere et al. 2016). Moreover, lignocellulosic biomasses are low-cost feedstocks and abundantly available globally with a continual supply (Nanda et al. 2016a; Okolie et al. 2020b). The production of biohydrogen from waste biomass has been reported extensively in the literature (Magnusson et al. 2008; Guo et al. 2010; Chen et al. 2012; Han et al. 2012; Moodley and Kana 2015; Kumar et al. 2017). Microbial communities utilize a wide variety of biomass resources for generating hydrogen. The types of microorganisms and feedstock have great roles in biohydrogen production. The biological processes for biohydrogen production include photolysis, dark fermentation, photo-fermentation and microbial electrolysis cells (Holladay et al. 2009).

Biohydrogen production through photo-fermentation is performed by utilizing photosynthetic bacteria through the enzyme nitrogenase system with the help of light energy as well as waste biomass. Purple non-sulfur bacteria help in the photo-fermentation process to utilize the reduced organic acids as a carbon source in the presence of solar light, thereby releasing molecular hydrogen with the help of a nitrogenase enzyme system (Basak and Das 2007). Some light-harvesting pigments like chlorophylls, phycobilins and carotenoids support electrons, protons and oxygen by utilizing sunlight via photo-fermentation (Figure 6.1). The nitrogenase enzyme system helps in the reaction of protons, electrons and nitrogen along with adenosine triphosphate (ATP) to produce hydrogen, ammonia, adenosine diphosphate (ADP) and inorganic phosphates (Pi) (Nanda et al. 2017b). The bacterial photosystem produces two electrons with four ATP molecules by utilizing light energy and biomass to generate hydrogen with the aid of the nitrogenase system.

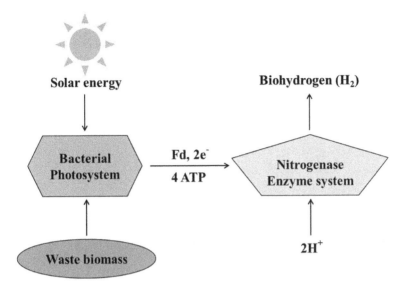

FIGURE 6.1   A simplified illustration of the photo-fermentation process for biohydrogen production

Dark fermentation is considered as a promising method for hydrogen production from waste biomass and specific microorganisms (Kumar et al. 2017; Łukajtis et al. 2018; Sarangi and Nanda 2020). During this method, a mixed gas containing H_2 and CO_2 is produced along with other gases like CH_4, CO and H_2S, which depends on the type of feedstock, microorganisms and process conditions (Datar et al. 2004; Najafpour et al. 2004; Kotsopoulos et al. 2006; Temudo et al. 2007). In dark fermentation, the bacterium converts organic substances like raw biomass, sugars and wastewater to hydrogen. Due to the complete absence of light, this process is regarded as dark fermentation. Moreover, dark fermentation is advantageous over photo-fermentation in requiring smaller bioreactors, less energy and low cost because of the absence of light energy to facilitate microbial growth. Some notable microorganisms like anaerobic bacteria such as *Bacillus* spp., *Enterobacter* spp. and *Clostridium* spp. are utilized for the conversion of cellulosic substrates to biohydrogen (Levin et al. 2004). During the dark fermentation, the bacterium converts glucose to pyruvic acid, thereby producing ATP through the glycolytic pathways. Furthermore, with the utilization of pyruvate ferredoxin oxidoreductase and hydrogenase, CO_2 and H_2 are produced from pyruvic acid (Figure 6.2). During dark fermentation, biohydrogen production generally depends on the degradation of pyruvate to acetyl-CoA and further to acetate, butyrate and ethanol.

Depending on the microorganisms employed and process conditions, different end products and by-products are generated from microbial biohydrogen production. The optimization of fermentation process parameters like sugar content, nutrients (including energy source and carbon source), hydrogen partial pressure, temperature, hydraulic retention time, pH, type of microorganisms used, inoculum pretreatment process, growth medium and cultural conditions can enhance the production of biohydrogen

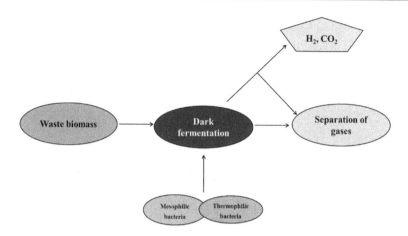

FIGURE 6.2 A simplified illustration of the dark fermentation process for biohydrogen production

(Ghimire et al. 2015). Besides, the bioreactor configuration, geometry and mode of operation can also affect biohydrogen production (Show et al. 2011).

Hydrogen production by microbial electrolysis cells is also regarded as a potential method for the utilization of energy as well as protons by microorganisms to convert the organic matter. Not only hydrogen but also various value-added chemicals are generated through this method, thereby establishing it as a promising platform for bioenergy utilization. Different value-added platform chemicals such as methane, formic acid and hydrogen peroxide are also released during this process with the potentials of wastewater treatment (Escapa et al. 2016). The process is similar to microbial fuel cells in having two compartments of anode and cathode separated by a proton exchange membrane (Nanda et al. 2017c; Bhatia et al. 2020). Protons (H^+) and electrons (e^-) are produced due to the oxidation of organic matter during this process. These are subsequently transferred to the cathode side through the membrane. Hydrogen is produced near the cathode by the reduction of electrons and protons in the presence of a catalyst (Logan et al. 2006).

6.3 CONCLUSIONS

Biohydrogen generation by microbial biomass is gaining attention considering its environmental friendliness and sustainability. Similarly, biohydrogen from waste biomass and organic residues are regarded as the most promising feedstocks owing to their renewability, abundance, low cost and polysaccharide composition. The exploration of microbial diversities along with the optimization of photo-fermentation and dark fermentation can maximize biohydrogen production. The integration of novel technologies in microbiology, chemical engineering and bioprocess engineering can create a road-map for efficient and sustainable biohydrogen production leading to energy security, economic sustainability and mitigation of greenhouse gas emissions.

REFERENCES

Basak, N., D. Das. 2007. The prospect of purple non-sulfur (PNS) photosynthetic bacteria for hydrogen production: the present state of the art. *World Journal of Microbiology and Biotechnology* 23:31–42.

Bhatia, L., P.K. Sarangi, S. Nanda. 2020. Current advancements in microbial fuel cell technologies. In: *Biorefinery of Alternative Resources: Targeting Green Fuels and Platform Chemicals*; eds. S. Nanda, D.V.N. Vo, P.K. Sarangi; 477–494. Singapore: Springer Nature.

Chen, C.C., Y.S. Chuang, C.Y. Lin, C.H. Lay, B. Sen. 2012. Thermophilic dark fermentation of untreated rice straw using mixed cultures for hydrogen production. *International Journal of Hydrogen Energy* 37:15540–15546.

Datar, R.P., R.M. Shenkman, B.G. Cateni, R.L. Huhnke, R.S. Lewis. 2004. Fermentation of biomass-generated producer gas to ethanol. *Biotechnology and Bioengineering* 86:587–594.

Escapa, A., R. Mateos, E.J.J. Martínez, J. Blanes. 2016. Microbial electrolysis cells: an emerging technology for wastewater treatment and energy recovery. From laboratory to pilot plant and beyond. *Renewable and Sustainable Energy Reviews* 55:942–956.

Fougere, D., S. Nanda, K. Clarke, J.A. Kozinski, K. Li. 2016. Effect of acidic pretreatment on the chemistry and distribution of lignin in aspen wood and wheat straw substrates. *Biomass & Bioenergy* 91:56–68.

Ghimire, A., L. Frunzo, F. Pirozzi, E. Trably, R. Escudie, P.N.L. Lens, G. Esposito. 2015. A review of dark fermentative biohydrogen production from organic biomass: process parameters and use of by-products. *Applied Energy* 144:73–95.

Guo, X.M., E. Trably, E. Latrille, H. Carrere, J.P. Steyer. 2010. Hydrogen production from agricultural waste by dark fermentation: a review. *International Journal of Hydrogen Energy* 35:10660–10673.

Han, H., L. Wei, B. Liu, H. Yang, J. Shen. 2012. Optimization of biohydrogen production from soybean straw using anaerobic mixed bacteria. *International Journal of Hydrogen Energy* 37:13200–13208.

Holladay, J.D., J. Hu, D.L. King, Y. Wang. 2009. An overview of hydrogen production technologies. *Catalysis Today* 139:244–260.

Huang, C.W., B.S. Nguyen, D.V.N. Vo, S. Nanda, V.H. Nguyen. 2020. Photocatalytic reforming for a sustainable hydrogen production over titania-based photocatalysts. In: *New Dimensions in Production and Utilization of Hydrogen*; ed. S. Nanda, D.V.N. Vo, P. Nguyen-Tri; Elsevier; pp. 191–213.

Kotsopoulos, T.A., R.J. Zeng, I. Angelidaki. 2006. Biohydrogen production in granular up flow anaerobic sludge blanket (UASB) reactors with mixed cultures under hyper-thermophilic temperature (70°C). *Biotechnology and Bioengineering* 94:296–302.

Kumar, G., P. Sivagurunathan, B. Sen, A. Mudhoo, G. Davila-Vazquez, G. Wang, S.H. Kim. 2017. Research and development perspectives of lignocellulose-based biohydrogen production. *International Biodeterioration & Biodegradation* 119:225–238.

Levin, D.B., L. Pitt, M. Love. 2004. Biohydrogen production: prospects and limitations to practical application. *International Journal of Hydrogen Energy* 29:173–185.

Logan, B.E., B. Hamelers, R. Rozendal, U. Schröder, J. Keller, S. Freguia, S.P. Aelterman, W. Verstraete, K. Rabaey. 2006. Microbial fuel cells: methodology and technology. *Environmental Science Technology* 40:5181–5192.

Łukajtis, R., I. Hołowacz, K. Kucharska, M. Glinka, P. Rybarczyk, A. Przyjazny, A. Kamiński. 2018. Hydrogen production from biomass using dark fermentation. *Renewable and Sustainable Energy Reviews* 91:665–694.

Magnusson, L., R. Islam, R. Sparling, D. Levin, N. Cicek. 2008. Direct hydrogen production from cellulosic waste materials with a single-step dark fermentation process. *International Journal of Hydrogen Energy* 33:5398–5403.

Moodley, P., E.B.G. Kana. 2015. Optimization of xylose and glucose production from sugarcane leaves (*Saccharum offinarum*) using hybrid pretreatment techniques and assessment for hydrogen generation at semi-pilot scale. *International Journal of Hydrogen Energy* 40:3859–3867.

Najafpour, G., H. Younesi, A.R. Mohamed. 2004. Effect of organic substrate on hydrogen practical application. *International Journal of Hydrogen Energy* 29:173–185.

Nanda, S., A.K. Dalai, F. Berruti, J.A. Kozinski. 2016a. Biochar as an exceptional bioresource for energy, agronomy, carbon sequestration, activated carbon and specialty materials. *Waste and Biomass Valorization* 7:201–235.

Nanda, S., A.K. Dalai, J.A. Kozinski. 2014a. Butanol and ethanol production from lignocellulosic feedstock: biomass pretreatment and bioconversion. *Energy Science & Engineering* 2:138–148.

Nanda, S., D. Golemi-Kotra, J.C. McDermott, A.K. Dalai, I. Gökalp, J.A. Kozinski. 2017a. Fermentative production of butanol: perspectives on synthetic biology. *New Biotechnology* 37:210–221.

Nanda, S., J. Maley, J.A. Kozinski, A.K. Dalai. 2015a. Physico-chemical evolution in lignocellulosic feedstocks during hydrothermal pretreatment and delignification. *Journal of Biobased Materials and Bioenergy* 9:295–308.

Nanda, S., J. Mohammad, S.N. Reddy, J.A. Kozinski, A.K. Dalai. 2014b. Pathways of lignocellulosic biomass conversion to renewable fuels. *Biomass Conversion and Biorefinery* 4:157–191.

Nanda, S., K. Li, N. Abatzoglou, A.K. Dalai, J.A. Kozinski. 2017b. Advancements and confinements in hydrogen production technologies. In: *Bioenergy Systems for the Future*; eds. F. Dalena, A. Basile, C. Rossi; 373–418. Cambridge: Woodhead Publishing, Elsevier.

Nanda, S., P. Mohanty, K.K. Pant, S. Naik, J.A. Kozinski, A.K. Dalai. 2013. Characterization of North American lignocellulosic biomass and biochars in terms of their candidacy for alternate renewable fuels. *Bioenergy Research* 6:663–677.

Nanda, S., R. Azargohar, A.K. Dalai, J.A. Kozinski. 2015b. An assessment on the sustainability of lignocellulosic biomass for biorefining. *Renewable and Sustainable Energy Reviews* 50:925–941.

Nanda, S., R. Rana, Y. Zheng, J.A. Kozinski, A.K. Dalai. 2017c. Insights on pathways for hydrogen generation from ethanol. *Sustainable Energy & Fuels* 1:1232–1245.

Nanda, S., S.N. Reddy, S.K. Mitra, J.A. Kozinski. 2016b. The progressive routes for carbon capture and sequestration. *Energy Science & Engineering* 4:99–122.

Okolie, J.A., R. Rana, S. Nanda, A.K. Dalai, J.A. Kozinski. 2019. Supercritical water gasification of biomass: a state-of-the-art review of process parameters, reaction mechanisms and catalysis. *Sustainable Energy & Fuels* 3:578–598.

Okolie, J.A., S. Nanda, A.K. Dalai, F. Berruti, J.A. Kozinski. 2020a. A review on subcritical and supercritical water gasification of biogenic, polymeric and petroleum wastes to hydrogen-rich synthesis gas. *Renewable and Sustainable Energy Reviews* 119:109546.

Okolie, J.A., S. Nanda, A.K. Dalai, J.A. Kozinski. 2020b. Chemistry and specialty industrial applications of lignocellulosic biomass. *Waste and Biomass Valorization*. Doi: 10.1007/s12649-020-01123-0.

Rana, R., S. Nanda, A. Maclennan, Y. Hu, J.A. Kozinski, A.K. Dalai. 2019. Comparative evaluation for catalytic gasification of petroleum coke and asphaltene in subcritical and supercritical water. *Journal of Energy Chemistry* 31:107–118.

Rana, R., S. Nanda, J.A. Kozinski, A.K. Dalai. 2018. Investigating the applicability of Athabasca bitumen as a feedstock for hydrogen production through catalytic supercritical water gasification. *Journal of Environmental Chemical Engineering* 6:182–189.

Rana, R., S. Nanda, S.N. Reddy, A.K. Dalai, J.A. Kozinski, I. Gökalp. 2020. Catalytic gasification of light and heavy gas oils in supercritical water. *Journal of the Energy Institute* 93: 2025–2032.

Reddy, S.N., S. Nanda, D.V.N. Vo, T.D. Nguyen, V.H. Nguyen, B. Abdullah, P. Nguyen-Tri. 2020. Hydrogen: fuel of the near future. In: *New Dimensions in Production and Utilization of Hydrogen*; ed. S. Nanda, D.V.N. Vo, P. Nguyen-Tri; Elsevier; pp. 1–20.

Sarangi, P.K., S. Nanda. 2018. Recent developments and challenges of acetone-butanol-ethanol fermentation. In: *Recent Advancements in Biofuels and Bioenergy Utilization*; eds. P.K. Sarangi, S. Nanda, P. Mohanty; 111–123. Singapore: Springer Nature.

Sarangi, P.K., S. Nanda. 2019. Recent advances in consolidated bioprocessing for microbe-assisted biofuel production. In: *Fuel Processing and Energy Utilization*; eds. S. Nanda, P.K. Sarangi, D.V.N. Vo; 141–157. Boca Raton, FL: CRC Press.

Sarangi, P.K., S. Nanda. 2020. Biohydrogen production through dark fermentation. *Chemical Engineering & Technology* 43:601–612.

Sarangi, P.K., S. Nanda, D.V.N. Vo. 2020. Technological advancements in the production and application of biomethanol. In: *Biorefinery of Alternative Resources: Targeting Green Fuels and Platform Chemicals*; eds. S. Nanda, D.V.N. Vo, P.K. Sarangi; 127–139. Singapore: Springer Nature.

Show, K.Y., D.J. Leeb, J.S. Chang. 2011. Bioreactor and process design for biohydrogen production. *Bioresource Technology* 102:8524–8533.

Siang, T.J., A.A. Jalil, M.N.N. Shafiqah, M.B. Bahari, H.D. Setiabudi, S.Z. Abidin, T.D. Nguyen, A. Abdulrahman, Q.V. Le, S. Nanda, D.V.N. Vo. 2020. Recent progress in ethanol steam reforming for hydrogen generation. In: *New Dimensions in Production and Utilization of Hydrogen*; ed. S. Nanda, D.V.N. Vo, P. Nguyen-Tri; Elsevier; pp. 57–80.

Temudo, M.F., R. Kleerebezem, M. van Loosdrecht. 2007. Influence of the pH on (open) mixed culture fermentation of glucose: a chemostat study. *Biotechnology and Bioengineering* 98:69–79.

Vakulchuk, R., I. Overland, D. Scholten. 2020. Renewable energy and geopolitics: a review. *Renewable and Sustainable Energy Reviews* 122:109547.

Conversion of Algal Biomass to Biofuels

7

7.1 INTRODUCTION

Fossil fuels and their derived products have played a pivotal role in advancements relating to automobiles, power generating sectors, infrastructure, urban development and human lifestyle since the industrial revolution. Nonetheless, the exploiting combustion of fossil fuels has led to the emission of greenhouse gases, particularly CO_2, to an alarming level (Rana et al. 2018; Rana et al. 2019; Rana et al. 2020). This increase in the concentration of CO_2 and other greenhouse gases has disturbed the balance between the solar radiation obtained by the earth and its reflection, thereby enhancing the heat retention ability of earth that results in global warming (Nanda et al. 2016e; Parakh et al. 2020). Besides global warming, other environmental concerns associated with fossil fuel usage are pollution, rising fuel prices and geopolitical issues between crude oil importing countries and exporting countries (Nanda et al. 2015a; Okolie et al. 2020b). To address these problems, research in the field of bioenergy and other renewable energy sources has been intensifying (Sarangi and Nanda 2018; Sarangi and Nanda 2019a; Sarangi and Nanda 2019b; Sarangi and Nanda 2020; Sarangi et al. 2020).

Several waste residues such as lignocellulosic biomass (e.g. agricultural crop residues, forestry refuse, energy crops and invasive crops), microalgae, food waste, municipal solid waste, sewage sludge, industrial effluents, livestock manure and other waste organic matter have the potential to be converted to biofuels, biochemicals and bioproducts with a low-carbon footprint or with carbon neutrality (Nanda et al. 2015b; Nanda et al. 2015c; Reddy et al. 2016; Nanda et al. 2016b; Nanda et al. 2016c; Nanda et al. 2016d; Gong et al. 2017a; Gong et al. 2017b; Nanda et al. 2017d; Nanda et al. 2018b; Nanda et al. 2019a; Nanda et al. 2019b; Singh et al. 2020; Okolie et al. 2020a; Okolie et al. 2020c; Okolie et al. 2020d). The biofuel products resulting from the thermochemical and biological conversion of the above-mentioned waste residues are bio-oil, biodiesel, bioethanol, biobutanol, biomethanol, biogas (biomethane), biohydrogen, synthesis gas and biochar (Nanda et al. 2014b; Nanda et al. 2016a; Nanda et al. 2017e; Okolie et al. 2019).

Microalgae has many promising potentials to produce biofuels, biochemicals and bioproducts (e.g. nutraceuticals, pharmaceuticals and cosmeceuticals) besides carbon sequestration and phycoremediation of wastewater (Reddy et al. 2018; Yadav et al. 2019). Microalgae are single phototrophic organisms in freshwater and marine environments. By utilizing sunlight, CO_2, water, organic matter and dissolved nutrients, algae can synthesize lipids, proteins, carbohydrates and pigments in their cells that can be extracted to produce various useful bioproducts (Koller et al. 2014). Microalgae find their potential in human and animal nutrition, whereas their extractives can also be used as a starting material in textile, pharmaceutical, cosmetics and food industries (Koller et al. 2014). Besides, algal biomass has gained promising applications in biofuel sectors for solving future energy problems acting as the third-generation biofuel feedstock to produce biofuels such as biodiesel, algal oil and bio-jet fuel (Reddy et al. 2018; Yadav et al. 2019). This chapter discusses the potential of algae for the next-generation biorefineries.

7.2 BIOPROSPECTING AND CULTIVATION OF ALGAE

Figure 7.1 shows the conversion of algal biomass to several value-added products. Algae biorefinery depends on the compositions of algal species such as lipids, carbohydrates and proteins, which serve as the precursors for biofuels and biochemicals (Laurens et al. 2017). Furthermore, the production of nutraceuticals, pharmaceuticals, pigments, vitamins, antioxidants can be obtained from algal biomass. Microalgae are also the important reservoirs of high-value nutrients, pigments, proteins, carbohydrates and lipid molecules (Ghosh et al. 2016). Algae are the known producers of various value-added compounds including extracellular products like exopolysaccharides, exoenzymes, etc. (Pierre et al. 2019). Besides biofuel production, algae can also be used for the phycoremediation of wastewater and industrial effluents in reducing the levels of hazardous chemicals and organic matter (Kumar et al. 2018). The promising option for the utilization of algae is for capturing the runoff fertilizers from different farms to lakes and water reservoirs.

Chlamydomonas reinhardtii and *Dunaliella salina* are some fast-growing species of algae that have been extensively explored along with various *Chlorella* sp. *Botryococcus braunii* can accumulate enormous quantities of lipids (Scott et al. 2010; Dragone et al. 2011). The lipid content of *Chlorella* sp. is very high (approximately 60–70%), making it quite popular among algal varieties. There are some geographical and cultural variations associated with algal biomass, which determine their lipid composition.

The growth of microalgae is dependent upon the type of cultivation system employed, availability of light, levels of CO_2 and O_2, temperature and availability of nutrients like nitrogen and phosphorus (Abdollahi and Dubljevic 2012; Li and Yang 2013). The three basic modes of algal cultivation are photoautotrophic, heterotrophic

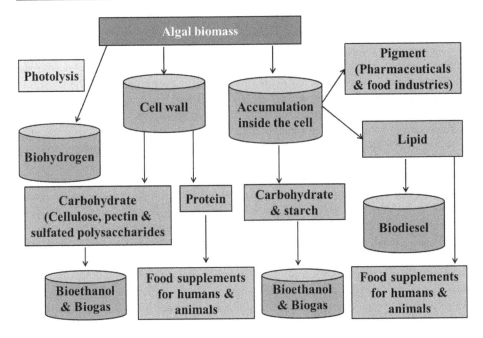

FIGURE 7.1   Biorefinery of algal biomass towards biofuels and platform chemicals

and mixotrophic. In photoautotrophic cultivation, algal growth using CO_2 is beneficial as an inexpensive process. On the other hand, the heterotrophic growth of algae consumes sugars and organic acids as the carbon sources in the absence of light. However, the mixotrophic mode uses both organic matter and CO_2 along with light for cultivating algae. The process of photosynthesis and carbon fixation by algae is much more efficient and faster than many terrestrial plants (Sayre 2010). Various harvesting processes such as centrifugation, filtration, flotation, sedimentation and flocculation along other downstream process technologies are used for the separation of algae and other bioproducts. Several flocculation techniques such as biological, magnetic, chemical, electro and auto-flocculation are used in this process.

The lipids composition of microalgal cells differs based on the algal variety and species, ingredients of the culture medium and environmental conditions along with growth factors such as temperature, complementation, luminous power and photoperiodicity (Brown 1991; Khoeyi et al. 2012; Procházková et al. 2014). At the optimal conditions, microalgae rapidly multiply but the accumulation of reserve substances like carbohydrates and lipids may not take place. During adverse conditions, it tends to rouse the gathering of pigments. Thus, regulations in experimental conditions are very crucial in the production of metabolites as the main products or by-products.

As mentioned earlier, the composition of the culture medium has a great role to play in the growth of algal cells and their proliferation. Diverse microalgal species differ in their nutritional necessities, although they can adapt to various supplementation conditions. Industrial and domestic wastewaters are also suitable as a culture medium

(Lv et al. 2017; Reyimu and Ozçimen 2017; Wu et al. 2017). Synthetic culture media with a well-defined composition can assess the response of algal cells to the alteration in concentration and modification of components. In certain situations, the product of interest can be obtained if the organism's metabolic pathway is stimulated after the supplementation of particular nutrients. The usage of cost-effective nutrient sources such as urea, human urine as well as glycerol have also been reported (Campos et al. 2014; Sengmee et al. 2017; Wu et al. 2017). The stirring process in microalgal cultures is important to reduce the sedimentation and homogenization of cell suspension. The stirrer is an essential part of a photobioreactor to enhance the cells to receive an equal quantity of light.

Microalgal growth is facilitated in either the existence or deficiency of light depending on the variety and strain. Algae can utilize both organic and inorganic carbon as the energy source. Photoautotrophic condition is widely employed for the cultivation of algae, which occurs in the presence of light. According to Kim et al. (2013), the intensity of light along with its color regulates the biomolecules developed and accumulated by the algal cells. Light is not mandatory for the occurrence of biochemical reactions in heterotrophic culture as organic carbon acts as the energy source. Light along with organic carbon sources are prerequisites for culturing micro-algae in mixotrophic and photoheterotrophic growth conditions. During mixotrophic culture, photosynthesis is performed by cells by using both organic as well as inorganic carbon. On the other hand, during the photoheterotrophic condition, light is needed by cells to use organic compounds (Chen et al. 2013). Certain photobioreactors can be employed using indoor artificial lighting for the growth of microalgal cells on a laboratory scale. Outdoor lighting is suitable for simulation studies either on a small or pilot scale to enhance productivity (Lu et al. 2015). These factors regulate the production and accumulation of lipids and carbohydrates in microalgae to estimate the yield of bio-crude oil, biohydrogen, biodiesel, bioethanol, biobutanol and other biofuels, biochemicals and bioproducts.

7.3 BIODIESEL FROM ALGAE

The lipid content of algae decides its potential to generate biofuels. Algae have huge potential as a feedstock for biodiesel production because of the following attributes (Chisti 2008; Mata et al. 2010):

    i. Enormous lipid content (20–50%).
    ii. Fast growth rates.
   iii. No competition to cultivable or fertile lands.
   iv. Capability to develop in rigorous conditions.
    v. Growth on low-cost substrates and wastewater effluents.
   vi. Ability to capture and fix CO_2 from flue gases.
  vii. Cost-effective and eco-friendly resource.

The emission of CO_2 from fossil fuels (e.g. diesel and gasoline) is a major environmental concern for which the focus has now gradually shifted towards renewable biofuels (e.g. biodiesel) (Reddy et al. 2018; Nayak et al. 2019; Nayak et al. 2020). Widjaja et al. (2009) stated that the accumulation of lipids as triacylglycerides takes place in microalgal cells when the environmental conditions become unfavorable (stress conditions) either in the form of nutrient deficit or the availability and intensity of light. The deficiency of nitrogen may significantly reduce the cell division as protein, pivotal for the formation of the cell wall, is produced in lesser amounts (Aremu et al. 2015). The production of biomass is also negatively affected when algae are deprived of phosphate. Similarly, significant reduction in the lipid concentration with the significant hike in unsaturated fatty acids concentration is seen (Praveenkumar et al. 2012). The cell development and lipid accumulation of lipids in microalgae are stimulated when an organic carbon source is supplied to the growth media.

The transesterification process or hydrogenolysis helps in the extraction of lipids from algae to yield biodiesel and aviation fuel (a derivative of kerosene grade alkane) (Bwapwa et al. 2017). By successive steps of methanolysis, triacylglycerols found in the algal oil can be broken down into diglycerides and monoglycerides. Using several acids, bases, metal catalysts, biocatalysts (e.g. enzymes) and supercritical fluids, the efficiency of transesterification can be enhanced (Reddy et al. 2018). Finally, three chief products such as fatty acid methyl esters (FAME), fatty acid ethyl esters (FAEE) and glycerol are generated from transesterification (Gong and Jiang 2011). If complete solubilization of triacylglycerols does not take place by the solvents, the extractive process is considered incompetent resulting in lower oil yields (Velasquez-Orta et al. 2013). Another aspect i.e. biomass drying temperature can also considerably affect the recovery of oil since the fatty acids are oxidized at high temperatures (Widjaja et al. 2009). Consequently, the oil extraction processes and techniques from microalgal cells can influence the yields of biodiesel. Lipid extraction becomes difficult when the water content of algal biomass is high. Therefore, it is necessary to dewater it by either centrifugation or filtration. The processes of culturing and dewatering are energy-intensive, which can add significantly to the overall process expenditures (Ríos et al. 2013).

7.4 BIOHYDROGEN FROM ALGAE

Biohydrogen is the most potent fuel that can substitute conventional fuels in the intermediate and extended terms (Reddy et al. 2014a; Reddy et al. 2014b; Nanda et al. 2017c; Reddy et al. 2020). Hydrogen is probably the cleanest fuel, energy vector and energy carrier because its combustion releases a significant amount of heat and water. The major limitations associated with the efficient usage of hydrogen lies in the fact that its production cost is high along with the difficulties associated with its safe storage and transportation (Khetkorn et al. 2017).

The main sources for hydrogen are fossil fuels, natural gas, organic biomass and water (Sarangi and Nanda 2020). Catalytic reforming, electrolysis, photolysis, thermochemical decomposition (e.g. gasification and pyrolysis), photoelectrochemical and biological systems (e.g. dark fermentation, photo-fermentation and microbial electrolytic cell) are the main routes for hydrogen production (Holladay et al. 2009; Nanda et al. 2017e; Singh et al. 2018; Shafiqah et al. 2020). Biohydrogen can be produced biologically involving certain microorganisms' photosynthetic biomachinery or by non-photosynthetic processes (Khetkorn et al. 2017). Metabolic reactions of microalgal cells also generate hydrogen. Water biophotolysis (direct or indirect) during photosynthesis generates hydrogen by green microalgae (Nanda et al. 2017c).

The photosystems I and II in algae capture sunlight during oxygenic photosynthesis, thereby mediating direct biophotolysis. During direct biophotolysis, breaking down of water molecules takes place, thus producing hydrogen with subsequent release of oxygen. The carbohydrate (starch) produced during the dark reaction generates hydrogen through the indirect biophotolysis process. The carbohydrate is produced biologically in the existence of water and adsorbed CO_2. Hence, H_2 and CO_2 are generated by the breakdown of carbohydrates. Due to high sensitivity to oxygen, hydrogenase enzyme works under the anaerobic condition to produce hydrogen, whereas oxygenic photosynthesis generates oxygen (Khetkorn et al. 2017). The conditions considered favorable for algae to undergo photoproduction of hydrogen are when the freshwater algae are deprived of sulfur and phosphorus and when seawater algae are deprived of phosphorus (Sengmee et al. 2017).

7.5 BIOETHANOL AND BIOBUTANOL FROM ALGAE

Bioethanol and biobutanol are the products of fermentation of sugars obtained from lignocellulosic materials and organic wastes (Nanda et al. 2014a; Nanda et al. 2017b). The production of bioethanol and biobutanol is treated as a green technology due to its energy efficiency and ecologically benign nature for its renewable precursors. The combustion of alcohol-based biofuels is also cleaner owing to negligible emissions of CO, hydrocarbons and particulate matter (Balat et al. 2008). While the production of bioethanol is mediated by *Saccharomyces cerevisiae*, biobutanol production is performed through *Clostridium*-facilitated acetone-butanol-ethanol (ABE) fermentation (Nanda et al. 2017a; Nanda et al. 2020). Both bioethanol and biobutanol can be blended with gasoline in flexible ratios for use in vehicles.

Microalgae cell composition is affluent in lipids along with proteins, whereas its carbohydrate amount is relatively low (Hernández et al. 2015). Hence, the production of biodiesel from algae is extensively studied when compared to that of bioethanol and biobutanol. However, there are some notable studies reported on bioethanol and biobutanol production from the hydrolysis, saccharification and fermentation of algal biomass using several bacterial and fungal species (Hirano et al. 1997; Ueno et al. 1998; Kim et al. 2011; Ellis et al. 2012). For bioethanol and biobutanol production from microalgae, it is vital to explore the strain that possesses a significant amount of polysaccharides in its cell wall along with the ability to accumulate starch (Dragone et al. 2011).

7.6 THERMOCHEMICAL AND HYDROTHERMAL CONVERSION OF ALGAL BIOMASS

The residual algal biomass after the lipid extraction can also either be saccharified to recover fermentable monosaccharides for the production of alcohol-based biofuels, biogas and biohydrogen or can be used for thermochemical conversion (e.g. gasification, pyrolysis, liquefaction and carbonization). The advancement of third-generation biofuels depends on the production of microalgal biomass. Microalgae biomass can be transformed into valuable products by thermochemical conversion (e.g. gasification, pyrolysis and liquefaction) and biochemical conversion (e.g. fermentation and anaerobic digestion). Through hydrothermal processing (e.g. gasification, liquefaction and carbonization) can result in hydrogen-rich syngas, bio-crude oil and hydrochar (Yadav et al. 2019). Moreover, hydrothermal processing technologies are found to be suitable for high-moisture containing algal biomass because it reduces the overall cost of biomass drying, as the reaction medium is water. Hydrothermal gasification and liquefaction involve the use of subcritical and supercritical water as the reaction medium, which act as green solvents to crack algal biomass to fuel products. Subcritical water is a fluid phase of water occurring below its critical points, whereas subcritical water occurs beyond its critical points (Reddy et al. 2014b). The critical temperature and critical pressure of water are 375°C and 22.1 MPa, respectively (Nanda et al. 2018c).

During pyrolysis, biomass undergoes thermal depolymerization at moderate to high temperatures under an inert atmosphere to produce bio-oil, biochar and gases (Azargohar et al. 2013; Mohanty et al. 2013; Nanda et al. 2013; Azargohar et al. 2014; Nanda et al. 2014d). The bio-oil can be catalytically upgraded to produce synthetic transportation fuels or serve as a precursor for numerous value-added platform chemicals (Nanda et al. 2014c). There is a wide array of value-added products such as Omega-3 fatty acids that can be obtained from microalgal oil, thus making it more economically justified for a sustainable bioresource (Gutiérrez et al. 2017). The hydrochar or bio-char obtained from hydrothermal and thermochemical processing of algae can be used for several applications in agronomy, biomedicine, carbon sequestration, adsorption of environmental pollutants, support of metal catalysts, production for activated carbon and other specialty materials (Nanda et al. 2016a; Nanda et al. 2018a).

7.7 CONCLUSIONS

Microalgae are considered as a sustainable and economical source for the production of biofuels, biochemicals and bioproducts (e.g. food supplements, pharmaceuticals, nutraceuticals and cosmeceuticals) having industry-wide importance. Having potential in biorefineries, extensive researches are being conducted for the recovery of value-added products. Algae are potential sources for carbon sequestration as they consume CO_2 for photosynthesis and release oxygen, while at the same time, fixing the carbon as lipids and polysaccharides within its cells. Moreover, algae can also be cultivated on diverse streams of wastewater and industrial effluents, thus leading to their phycoremediation and environmental scrubbing. A wide array of co-products can be generated along with algal biofuels to support biorefineries and making them profitable and environmentally sustainable. The optimization of the cultivation process for maximum lipid and biofuel recovery from algae is highly essential. The implementation of genetic engineering strategies and biotechnological tools can produce high-yielding varieties of algae.

REFERENCES

Abdollahi, J., S. Dubljevic. 2012. Lipid production optimization and optimal control of heterotrophic microalgae fed-batch bioreactor. *Chemical Engineering Science* 84:619–627.

Aremu, A.O., M. Neményi, W.A. Stirk, V. Ördög, J. van Staden. 2015. Manipulation of nitrogen levels and mode of cultivation are viable methods to improve the lipid, fatty acids, phytochemical content, and bioactivities in *Chlorella minutissima*. *Journal of Phycology* 51:659–669.

Azargohar, R., S. Nanda, B.V.S.K. Rao, A.K. Dalai. 2013. Slow pyrolysis of deoiled Canola meal: product yields and characterization. *Energy & Fuels* 27:5268–5279.

Azargohar, R., S. Nanda, J.A. Kozinski, A.K. Dalai, R. Sutarto. 2014. Effects of temperature on the physicochemical characteristics of fast pyrolysis bio-chars derived from Canadian waste biomass. *Fuel* 125:90–100.

Balat, M., H. Balat, C. Oz. 2008. Progress in bioethanol processing. *Progress Energy Combustion Science* 34:551–573.

Brown, M.R. 1991. The amino-acid and sugar composition of 16 species of microalgae used in mariculture. *Journal of Experimental Marine Biology and Ecology* 145:79–99.

Bwapwa, J.K., A. Anandraj, C. Trois. 2017. Possibilities for conversion of microalgae oil into aviation fuel: a review. *Renewable and Sustainable Energy Reviews* 80:1345–1354.

Campos, H., W.J. Boeing, B.N. Dungan, T. Schaub. 2014. Cultivating the marine microalga *Nannochloropsis salina* under various nitrogen sources: effect on biovolume yields, lipid content and composition, and invasive organisms. *Biomass and Bioenergy* 66:301–307.

Chen, C.Y., X.Q. Zhao, H.W. Yen, S.H. Ho, C.L. Cheng, D.J. Lee, F.W. Bai, J.S. Chang. 2013. Microalgae-based carbohydrates for biofuel production. *Biochemical Engineering Journal* 78:1–10.

Chisti, Y. 2008. Biodiesel from microalgae beats bioethanol. *Trends in Biotechnology* 26:126–131.

Dragone, G., B.D. Fernandes, A.P. Abreu, A.A. Vicente, J.A. Teixeira. 2011. Nutrient limitation as a strategy for increasing starch accumulation in microalgae. *Applied Energy* 88:3331–3335.

Ellis, J.T., N.N. Hengge, R.C. Sims, C.D. Miller. 2012. Acetone, butanol, and ethanol production from wastewater algae. *Bioresource Technology* 111:491–495.

Ghosh, A., S. Khanra, M. Mondal, G. Halder, O.N. Tiwari, S. Saini, T.K. Bhowmick, K. Gayen. 2016. Progress toward isolation of strains and genetically engineered strains of microalgae for production of biofuel and other value added chemicals: a review. *Energy Conversion and Management* 113:104–118.

Gong, M., S. Nanda, H.N. Hunter, W. Zhu, A.K. Dalai, J.A. Kozinski. 2017a. Lewis acid catalyzed gasification of humic acid in supercritical water. *Catalysis Today* 291:13–23.

Gong, M., S. Nanda, M.J. Romero, W. Zhu, J.A. Kozinski. 2017b. Subcritical and supercritical water gasification of humic acid as a model compound of humic substances in sewage sludge. *The Journal of Supercritical Fluids* 119:130–138.

Gong, Y., M. Jiang. 2011. Biodiesel production with microalgae as feedstock: from strains to biodiesel. *Biotechnology Letters* 33:1269–1284.

Gutiérrez, C.D.B., D.L.R. Serna, C.A.C. Alzate. 2017. A comprehensive review on the implementation of the biorefinery concept in biodiesel production plants. *Biofuel Research Journal* 15:691–703.

Hernández, D., B. Riaño, M. Coca, M.C. García-González. 2015. Saccharification of carbohydrates in microalgal biomass by physical, chemical and enzymatic pre-treatments as a previous step for bioethanol production. *Chemical Engineering Journal* 262:939–945.

Hirano, A., R. Ueda, S. Hirayama, Y. Ogushi. 1997. CO_2 fixation and ethanol production with microalgal photosynthesis and intracellular anaerobic fermentation. *Energy* 22:137–142.

Holladay, J.D., J. Hu, D.L. King, Y. Wang. 2009. An overview of hydrogen production technologies. *Catalysis Today* 139:244–260.

Khetkorn, W., R.P. Rastogi, A. Incharoensakdi, P. Lindblad, D. Madamwar, A. Pandey, C. Larrocheg. 2017. Microalgal hydrogen production—a review. *Bioresource Technology* 243:1194–1206.

Khoeyi, Z.A., J. Seyfabadi, Z. Ramezanpour. 2012. Effect of light intensity and photoperiod on biomass and fatty acid composition of the microalgae, *Chlorella vulgaris*. *Aquaculture International* 20:41–49.

Kim, N.J., H. Li, K. Jung, H.N. Chang, P.C. Lee. 2011. Ethanol production from marine algal hydrolysates using *Escherichia coli* KO11. *Bioresource Technology* 102:7466–7469.

Kim, T.H., Y. Lee, S.H. Han, S.J. Hwang. 2013. The effects of wavelength and wavelength mixing ratios on microalgae growth and nitrogen, phosphorus removal using *Scenedesmus* sp. for wastewater treatment. *Bioresource Technology* 130:75–80.

Koller, M., A. Muhr, G. Braunegg. 2014. Microalgae as versatile cellular factories for valued products. *Algal Research* 6:52–63.

Kumar, P.K., S.V. Krishna, K. Verma, K. Pooja, D. Bhagawan, V. Himabindu. 2018. Phycoremediation of sewage wastewater and industrial flue gases for biomass generation from microalgae. *South African Journal of Chemical Engineering* 25:133–146.

Laurens, L.M.L., J. Markham, D.W. Templeton, E.D. Christensen, S. Van Wychen, E.W. Vadelius, M. Chen-Glasser, T. Dong, R. Davis, P.T. Pienkos. 2017. Development of algae biorefinery concepts for biofuels and bioproducts; a perspective on process-compatible products and their impact on cost-reduction. *Energy Environment Science* 10:1716–1738.

Li, X., N. Yang. 2013. Modeling the light distribution in air lift photobioreactors under simultaneous external and internal illumination using the two-flux model. *Chemical Engineering Science* 88:16–22.

Lu, W., Z. Wang, X. Wang, Z. Yuan. 2015. Cultivation of *Chlorella* sp. using raw dairy waste water for nutrient removal and biodiesel production: characteristics comparison of indoor bench-scale and outdoor pilot-scale cultures. *Bioresource Technology* 192:382–388.

Lv, J., J. Guo, J. Feng, Q. Liu, S. Xie. 2017. Effect of sulfate ions on growth and pollutants removal of self-flocculating microalga *Chlorococcum* sp. GD in synthetic municipal wastewater. *Bioresource Technology* 234:289–296.

Mata, T.M., A.A. Martins, N.S. Caetano. 2010. Microalgae for biodiesel production and other applications: a review. *Renewable and Sustainable Energy Reviews* 14:217–232.

Mohanty, P., S. Nanda, K.K. Pant, S. Naik, J.A. Kozinski, A.K. Dalai. 2013. Evaluation of the physiochemical development of biochars obtained from pyrolysis of wheat straw, timothy grass and pinewood: effects of heating rate. *Journal of Analytical and Applied Pyrolysis* 104:485–493.

Nanda, S., A.K. Dalai, F. Berruti, J.A. Kozinski. 2016a. Biochar as an exceptional bioresource for energy, agronomy, carbon sequestration, activated carbon and specialty materials. *Waste and Biomass Valorization* 7:201–235.

Nanda, S., A.K. Dalai, I. Gökalp, J.A. Kozinski. 2016b. Valorization of horse manure through catalytic supercritical water gasification. *Waste Management* 52:147–158.

Nanda, S., A.K. Dalai, J.A. Kozinski. 2014a. Butanol and ethanol production from lignocellulosic feedstock: biomass pretreatment and bioconversion. *Energy Science & Engineering* 2:138–148.

Nanda, S., A.K. Dalai, J.A. Kozinski. 2016c. Supercritical water gasification of timothy grass as an energy crop in the presence of alkali carbonate and hydroxide catalysts. *Biomass & Bioenergy* 95:378–387.

Nanda, S., A.K. Dalai, J.A. Kozinski. 2017a. Butanol from renewable biomass: highlights on downstream processing and recovery techniques. In: *Sustainable Utilization of Natural Resources*; eds. P. Mondal, A.K. Dalai; 187–211. Boca Raton, FL: CRC Press.

Nanda, S., A.K. Dalai, K.K. Pant, I. Gökalp, J.A. Kozinski. 2018a. An appraisal on biochar functionality and utility in agronomy. In: *Bioenergy and Biofuels*; ed. O. Konur; 389–409. Boca Raton, FL: CRC Press.

Nanda, S., D. Golemi-Kotra, J.C. McDermott, A.K. Dalai, I. Gökalp, J.A. Kozinski. 2017b. Fermentative production of butanol: perspectives on synthetic biology. *New Biotechnology* 37:210–221.

Nanda, S., J. Isen, A.K. Dalai, J.A. Kozinski. 2016d. Gasification of fruit wastes and agro-food residues in supercritical water. *Energy Conversion and Management* 110:296–306.

Nanda, S., J. Mohammad, S.N. Reddy, J.A. Kozinski, A.K. Dalai. 2014b. Pathways of lignocellulosic biomass conversion to renewable fuels. *Biomass Conversion and Biorefinery* 4:157–191.

Nanda, S., K. Li, N. Abatzoglou, A.K. Dalai, J.A. Kozinski. 2017c. Advancements and confinements in hydrogen production technologies. In: *Bioenergy Systems for the Future*; eds. F. Dalena, A. Basile, C. Rossi; 373–418. Cambridge: Woodhead Publishing, Elsevier.

Nanda, S., M. Gong, H.N. Hunter, A.K. Dalai, I. Gökalp, J.A. Kozinski. 2017d. An assessment of pinecone gasification in subcritical, near-critical and supercritical water. *Fuel Processing Technology* 168:84–96.

Nanda, S., P. Mohanty, J.A. Kozinski, A.K. Dalai. 2014c. Physico-chemical properties of bio-oils from pyrolysis of lignocellulosic biomass with high and slow heating rate. *Energy and Environment Research* 4:21–32.

Nanda, S., P. Mohanty, K.K. Pant, S. Naik, J.A. Kozinski, A.K. Dalai. 2013. Characterization of North American lignocellulosic biomass and biochars in terms of their candidacy for alternate renewable fuels. *Bioenergy Research* 6:663–677.

Nanda, S., R. Azargohar, A.K. Dalai, J.A. Kozinski. 2015a. An assessment on the sustainability of lignocellulosic biomass for biorefining. *Renewable and Sustainable Energy Reviews* 50:925–941.

Nanda, S., R. Azargohar, J.A. Kozinski, A.K. Dalai. 2014d. Characteristic studies on the pyrolysis products from hydrolyzed Canadian lignocellulosic feedstocks. *Bioenergy Research* 7:174–191.

Nanda, S., R. Rana, D.V.N. Vo, P.K. Sarangi, T.D. Nguyen, A.K. Dalai, J.A. Kozinski. 2020. A spotlight on butanol and propanol as next-generation synthetic fuels. In: *Biorefinery of Alternative Resources: Targeting Green Fuels and Platform Chemicals*; eds. S. Nanda, D.V.N. Vo, P.K. Sarangi; 105–126. Singapore: Springer Nature.

Nanda, S., R. Rana, H.N. Hunter, Z. Fang, A.K. Dalai, J.A. Kozinski. 2019a. Hydrothermal catalytic processing of waste cooking oil for hydrogen-rich syngas production. *Chemical Engineering Science* 195:935–945.

Nanda, S., R. Rana, Y. Zheng, J.A. Kozinski, A.K. Dalai. 2017e. Insights on pathways for hydrogen generation from ethanol. *Sustainable Energy & Fuels* 1:1232–1245.

Nanda, S., S.N. Reddy, D.V.N. Vo, B.N. Sahoo, J.A. Kozinski. 2018b. Catalytic gasification of wheat straw in hot compressed (subcritical and supercritical) water for hydrogen production. *Energy Science & Engineering* 6:448–459.

Nanda, S., S.N. Reddy, H.N. Hunter, A.K. Dalai, J.A. Kozinski. 2015b. Supercritical water gasification of fructose as a model compound for waste fruits and vegetables. *The Journal of Supercritical Fluids* 104:112–121.

Nanda, S., S.N. Reddy, H.N. Hunter, D.V.N. Vo, J.A. Kozinski, I. Gökalp. 2019b. Catalytic subcritical and supercritical water gasification as a resource recovery approach from waste tires for hydrogen-rich syngas production. *The Journal of Supercritical Fluids* 154:104627.

Nanda, S., S.N. Reddy, H.N. Hunter, I.S. Butler, J.A. Kozinski. 2015c. Supercritical water gasification of lactose as a model compound for valorization of dairy industry effluents. *Industrial & Engineering Chemistry Research* 54:9296–9306.

Nanda, S., S.N. Reddy, S.K. Mitra, J.A. Kozinski. 2016e. The progressive routes for carbon capture and sequestration. *Energy Science & Engineering* 4:99–122.

Nanda, S., S.N. Reddy, Z. Fang, A.K. Dalai, J.A. Kozinski. 2018c. Hydrothermal events occurring during gasification in supercritical water. In: *Supercritical and Other High-Pressure Solvent Systems: For Extraction, Reaction and Material Processing*; eds. A.J. Hunt, T.M. Attard; 560–587. London: Royal Society of Chemistry.

Nayak, S.K., B. Nayak, P.C. Mishra, M.M. Noor, S. Nanda. 2019. Effects of biodiesel blends and producer gas flow on overall performance of a turbocharged direct injection dual-fuel engine. *Energy Sources, Part A: Recovery, Utilization, and Environmental Effects*. Doi: 10.1080/15567036.2019.1694101.

Nayak, S.K., P.C. Mishra, S. Nanda, B. Nayak, M.M. Noor. 2020. Opportunities for biodiesel compatibility as a modern combustion engine fuel. In: *Biorefinery of Alternative Resources: Targeting Green Fuels and Platform Chemicals*; eds. S. Nanda, D.V.N. Vo, P.K. Sarangi; 457–476. Singapore: Springer Nature.

Okolie, J.A., R. Rana, S. Nanda, A.K. Dalai, J.A. Kozinski. 2019. Supercritical water gasification of biomass: a state-of-the-art review of process parameters, reaction mechanisms and catalysis. *Sustainable Energy & Fuels* 3:578–598.

Okolie, J.A., S. Nanda, A.K. Dalai, F. Berruti, J.A. Kozinski. 2020a. A review on subcritical and supercritical water gasification of biogenic, polymeric and petroleum wastes to hydrogen-rich synthesis gas. *Renewable and Sustainable Energy Reviews* 119:109546.

Okolie, J.A., S. Nanda, A.K. Dalai, J.A. Kozinski. 2020b. Chemistry and specialty industrial applications of lignocellulosic biomass. *Waste and Biomass Valorization*. Doi: 10.1007/s12649-020-01123-0.

Okolie, J.A., S. Nanda, A.K. Dalai, J.A. Kozinski. 2020c. Hydrothermal gasification of soybean straw and flax straw for hydrogen-rich syngas production: experimental and thermodynamic modeling. *Energy Conversion and Management* 208:112545.

Okolie, J.A., S. Nanda, A.K. Dalai, J.A. Kozinski. 2020d. Optimization and modeling of process parameters during hydrothermal gasification of biomass model compounds to generate hydrogen-rich gas products. *International Journal of Hydrogen Energy* 45:18275–18288.

Parakh, P.D., S. Nanda, J.A. Kozinski. 2020. Eco-friendly transformation of waste biomass to biofuels. *Current Biochemical Engineering* 6:120–134.

Pierre, G., C. Delattre, P. Dubessay, S. Jubeau, C. Vialleix, J.P. Cadoret, I. Probert, P. Michaud. 2019. What is in store for EPS microalgae in the next decade? *Molecules* 24:4296.

Praveenkumar, R., K. Shameera, G. Mahalakshmi, M.A. Akbarsha, N. Thajuddin. 2012. Influence of nutrient deprivations on lipid accumulation in a dominant indigenous microalga *Chlorella* sp., BUM11008: evaluation for biodiesel production. *Biomass and Bioenergy* 37:60–66.

Procházková, G., I. Brányiková, V. Zachleder, T. Brányik. 2014. Effect of nutrient supply status on biomass composition of eukaryotic green microalgae. *Journal of Applied Phycology* 26:1359–1377.

Rana, R., S. Nanda, A. Maclennan, Y. Hu, J.A. Kozinski, A.K. Dalai. 2019. Comparative evaluation for catalytic gasification of petroleum coke and asphaltene in subcritical and supercritical water. *Journal of Energy Chemistry* 31:107–118.

Rana, R., S. Nanda, J.A. Kozinski, A.K. Dalai. 2018. Investigating the applicability of Athabasca bitumen as a feedstock for hydrogen production through catalytic supercritical water gasification. *Journal of Environmental Chemical Engineering* 6:182–189.

Rana, R., S. Nanda, S.N. Reddy, A.K. Dalai, J.A. Kozinski, I. Gökalp. 2020. Catalytic gasification of light and heavy gas oils in supercritical water. *Journal of the Energy Institute* 93, 2025–2032.

Reddy, S.N., N. Ding, S. Nanda, A.K. Dalai, J.A. Kozinski. 2014a. Supercritical water gasification of biomass in diamond anvil cells and fluidized beds. *Biofuels, Bioproducts and Biorefining* 8:728–737.

Reddy, S.N., S. Nanda, A.K. Dalai, J.A. Kozinski. 2014b. Supercritical water gasification of biomass for hydrogen production. *International Journal of Hydrogen Energy* 39:6912–6926.

Reddy, S.N., S. Nanda, D.V.N. Vo, T.D. Nguyen, V.H. Nguyen, B. Abdullah, P. Nguyen-Tri. 2020. Hydrogen: fuel of the near future. In: *New Dimensions in Production and Utilization of Hydrogen*; eds. S. Nanda, D.V.N. Vo, P. Nguyen-Tri; Elsevier, pp. 1–20.

Reddy, S.N., S. Nanda, J.A. Kozinski. 2016. Supercritical water gasification of glycerol and methanol mixtures as model waste residues from biodiesel refinery. *Chemical Engineering Research and Design* 113:17–27.

Reddy, S.N., S. Nanda, P.K. Sarangi. 2018. Applications of supercritical fluids for biodiesel production. In: *Recent Advancements in Biofuels and Bioenergy Utilization*; eds. P.K. Sarangi, S. Nanda, P. Mohanty; 261–284. Singapore: Springer Nature.

Reyimu, Z., D. Ozçimen. 2017. Batch cultivation of marine microalgae *Nannochloropsis oculata* and *Tetraselmis suecica* in treated municipal wastewater toward bioethanol production. *Journal of Cleaner Production* 150:40–46.

Ríos, S.D., C.M. Torres, C. Torras, J. Salvadó, J.M. Mateo-Sanz, L. Jiménez. 2013. Microalgae-based biodiesel: economic analysis of downstream process realistic scenarios. *Bioresource Technology* 136:617–625.

Sarangi, P.K., S. Nanda. 2018. Recent developments and challenges of acetone-butanol-ethanol fermentation. In: *Recent Advancements in Biofuels and Bioenergy Utilization*; eds. P.K. Sarangi, S. Nanda, P. Mohanty; 111–123. Singapore: Springer Nature.

Sarangi, P.K., S. Nanda. 2019a. Bioconversion of agro-wastes into phenolic flavour compounds. In: *Biotechnology for Sustainable Energy and Products*; eds. P.K. Sarangi, S. Nanda; 266–284. New Delhi: I.K. International Publishing House Pvt. Ltd.

Sarangi, P.K., S. Nanda. 2019b. Recent advances in consolidated bioprocessing for microbe-assisted biofuel production. In: *Fuel Processing and Energy Utilization*; eds. S. Nanda, P.K. Sarangi, D.V.N. Vo; 141–157. Boca Raton, FL: CRC Press.

Sarangi, P.K., S. Nanda. 2020. Biohydrogen production through dark fermentation. *Chemical Engineering & Technology* 43:601–612.

Sarangi, P.K., S. Nanda, D.V.N. Vo. 2020. Technological advancements in the production and application of biomethanol. In: *Biorefinery of Alternative Resources: Targeting Green Fuels and Platform Chemicals*; eds. S. Nanda, D.V.N. Vo, P.K. Sarangi; 127–139. Singapore: Springer Nature.

Sayre, R. 2010. Microalgae: the potential for carbon capture. *BioScience* 60:722–727.

Scott, S.A., M.P. Davey, J.S. Dennis, I. Horst, C.J. Howe, D.J. Lea-Smith, A.G. Smith. 2010. Biodiesel from algae: challenges and prospects. *Current Opinion in Biotechnology* 21:277–286.

Sengmee, D., B. Cheirsilp, T.T. Suksaroge, P. Prasertsan. 2017. Biophotolysis-based hydrogen and lipid production by oleaginous microalgae using crude glycerol as exogenous carbon source. *International Journal of Hydrogen Energy* 42:1970–1976.

Shafiqah, M.N.N., H.N. Tran, T.D. Nguyen, P.T.T. Phuong, B. Abdullah, S.S. Lam, P. Nguyen-Tri, R. Kumar, S. Nanda, D.V.N. Vo. 2020. Ethanol CO_2 reforming on La_2O_3 and CeO_2-promoted Cu/Al_2O_3 catalysts for enhanced hydrogen production. *International Journal of Hydrogen Energy* 45:18398–18410.

Singh, A., S. Nanda, F. Berruti. 2020. A review of thermochemical and biochemical conversion of *Miscanthus* to biofuels. In: *Biorefinery of Alternative Resources: Targeting Green Fuels and Platform Chemicals*; eds. S. Nanda, D.V.N. Vo, P.K. Sarangi; 195–220. Singapore: Springer Nature.

Singh, S., R. Kumar, H.D. Setiabudi, S. Nanda, D.V.N. Vo. 2018. Advanced synthesis strategies of mesoporous SBA-15 supported catalysts for catalytic reforming applications: a state-of-the-art review. *Applied Catalysis A: General* 559:57–74.

Ueno, Y., N. Kurano, S. Miyachi. 1998. Ethanol production by dark fermentation in the marine green alga, *Chlorococcum littorale*. *Journal of Fermentation and Bioengineering* 86:38–43.

Velasquez-Orta, S.B., J.G.M. Lee, A.P. Harvey. 2013. Evaluation of FAME production from wet marine and freshwater microalgae by in situ transesterification. *Biochemical Engineering Journal* 76:83–89.

Widjaja, A., C.C. Chien, Y.H. Ju. 2009. Study of increasing lipid production from fresh water microalgae *Chlorella vulgaris*. *Journal of the Taiwan Institute of Chemical Engineers* 40:13–20.

Wu, J.Y., C.H. Lay, C.C. Chen, S.Y. Wu. 2017. Lipid accumulating microalgae cultivation in textile wastewater: environmental parameters optimization. *Journal of the Taiwan Institute of Chemical Engineers* 79:1–6.

Yadav, P., S.N. Reddy, S. Nanda. 2019. Cultivation and conversion of algae for wastewater treatment and biofuel production. In: *Fuel Processing and Energy Utilization*; eds. S. Nanda, P.K. Sarangi, D.V.N. Vo; 159–175. Florida, USA: CRC Press.

Perspectives on Microbial Fuel Cells

8

8.1 INTRODUCTION

The rapid industrial development and urban expansion have resulted in concerns relating to threatened energy security, environmental pollution, global warming and climate change both in the developing and developed nations (Nanda et al. 2015a; Nanda et al. 2016e; Nanda et al. 2017c). Non-renewable sources of energy have quenched the global energy requirement in the manufacturing, transportation, automobile as well as combined heat and power sectors since the industrial revolution (Rana et al. 2018; Rana et al. 2019; Rana et al. 2020). Currently, the world needs further exploration of alternative and renewable energy technologies to suffice the increasing energy demands while synergizing solutions for the above-mentioned environmental concerns (Okolie et al. 2019; Okolie et al. 2020a; Okolie et al. 2020b).

Some renewable sources of energy that are found to be promising are geothermal, wind, solar, tidal, algae, lignocellulosic biomass (e.g. agricultural crop refuse, forestry residues, dedicated energy crops and invasive crops) and organic waste materials (e.g. sewage sludge, municipal solid waste, waste tires, livestock manure, industrial effluents, food waste, etc.) (Nanda et al. 2015b; Nanda et al. 2016a; Nanda et al. 2016b; Nanda et al. 2016c; Reddy et al. 2016;Gong et al. 2017a; Gong et al. 2017b; Nanda et al. 2017b; Parakh et al. 2020; Singh et al. 2020). These waste residues can be transformed to some advanced biofuels such as bio-oil, biodiesel, bioethanol, biobutanol, biogas, biohydrogen, syngas, etc. through a wide variety of thermochemical and biological conversion technologies (Nanda et al. 2014; Nanda et al. 2016d; Sarangi and Nanda 2018; Sarangi and Nanda 2019; Sarangi and Nanda 2020; Sarangi et al. 2020). Besides, fuel cells are also some recent options being explored as a renewable source of energy using microorganisms and waste organic matter (Bhatia et al. 2020). Major merit associated with the employment of fuel cells is zero-emission of harmful greenhouse gases (viz. SO_x, NO_x, CO_2 and CO) (Rahimnejad et al. 2015; Bhatia et al. 2020). This chapter provides an overview of microbial fuel cells for the production of bioelectricity.

8.2 VARIATIONS OF MICROBIAL FUEL CELLS

Two forms of biological fuel cells are utilized such as enzymatic fuel cell (EFC) and microbial fuel cell (MFC). Selective enzymes are involved in EFC for the redox reaction, whereas MFC involves microorganisms (or electroactive microorganisms) in an anaerobic anode compartment to generate electricity from organic compounds. The potential of electroactive bacteria (EAB) has been recognized to oxidize a series of organic matter or pollutant serving them with carbon for their metabolism. During this process, EAB transfer the produced electron to anodes (Shen et al. 2014). Hence, MFCs are renewable devices to transform chemical energy into electricity by employing the anaerobic metabolic machinery of EAB.

The MFC technology is highly beneficial for having the capacity of converting waste organic matter and pollutants into electricity using various microorganisms and their enzymes. The MFCs differ from conventional fuel cells in many aspects. For example, MFCs operate at ambient temperature range (approximately 15–45°C), which involves biotic electrocatalyst at the anodic side and neutral pH condition. The MFCs can utilize complex biomass and has less environmental adverse impacts when compared to the traditional fuel cells (He et al. 2005; Larrosa-Guerrero et al. 2010; Borole et al. 2011; Tremouli et al. 2016).

Based on the assembly of the cathode and anode chambers, MFCs can be classified as single-chambered or double-chambered (Figure 8.1). For enhancing the efficiency of MFCs, several modifications have been adopted in their basic design and structure. The electrons generated in the anode chamber by the oxidation of the substrate move towards the cathode, which is managed with the help of a mediator or without a mediator.

Depending on the movement of electrons from the EAB to the anode, MFCs can be classified into two types such as MFCs with mediator and MFCs without a mediator (Pant et al. 2010). In the first category, the MFCs employ mediators that are supplemented to the system. In the latter category, certain microorganisms aid in the transfer of electrons via conductive pili or via cytochromes associated with their membrane and are electrochemically active. In certain instances, the redox-mediating molecules secreted by a microorganism also govern this mechanism. Some metal-reducing bacteria such as *Geobacter metallireducens*, *Rhodoferax ferrireducens*, *Shewanella putrefaciens*, *Clostridium butyric* and *Aeromonas hydrophila* exhibit this phenomenon of mediator-less electron transport in MFCs. Soluble redox shuttles play an important role in the power generation when MFCs involve a conglomerate of *Alcaligenes faecalis*, *Enterococcus faecium* and *Pseudomonas aeruginosa* secreting these redox shuttles. The added mediators can sometimes pose the problems of toxicity and instability to limit their applications in MFCs. The employment of microbial-generated native electron shuttles can resolve this issue. It is interesting to employ the secondary metabolites as redox mediators for MFC applications, as their in-situ production curtails the need for adding exogenous redox shuttles to transfer the electrons.

Sediment-type microbial fuel cell (SMFC) is a form of MFC in which anode is submerged in the anaerobic sediment comprising of detritus organic material of plant and

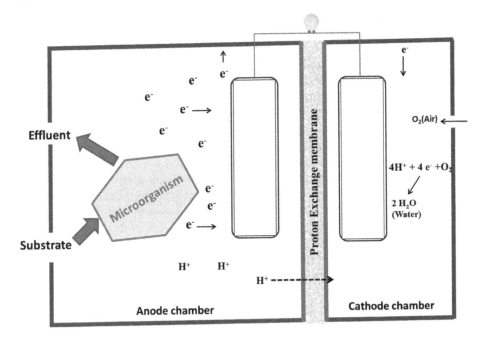

FIGURE 8.1   A typical double-chambered microbial fuel cell

animal and human origin. An electric circuit joins this anode and a cathode electrode dangled in superimposing water (Xu et al. 2015). The feasibility of this design depends on the concept that exoelectrogens can utilize the organic carbon found in these sediments and liberate electrons that are transported outside the cells (Holmes et al. 2004).

8.3 NEW DEVELOPMENTS IN MICROBIAL FUEL CELLS

The implementation of MFC technology aids in the generation of clean energy while providing one of the best platforms for bioremediation of pollutants. The microbial consortia are employed in the anode chamber of the MFC where they oxide the wastewater or the pollutant to generate the electrons and protons. Theses microorganisms utilize the electrons by the electron transfer chain for their metabolism after which they travel to the cathode chamber completing the circuit. The protons produced in the anode chamber during the oxidation process also travel via the proton exchange membrane (PEM) to reach the cathode. The cathode chamber is oxygenated where the electrons and protons reduce the oxygen to produce water molecules, thereby completing the charge balance while converting the chemical energy to electrical energy (Osman et al.

2010). The extracellular electron transport mechanism of EAB supports the transport of electron, thereby generating voltage (Wang et al. 2014). A study has shown the use of human urine as a source of energy in MFCs (Ieropoulos et al. 2013).

Different mechanisms are adopted for transferring electrons to the solid electrodes. Exoelectrogens in MFCs aid in transferring electrons. For example, pyocyanin or riboflavin is a mediator secreted by *Pseudomonas* and *Shewanella* that accelerates the transfer of electrons. *Pseudomonas aeruginosa* is also a potent secretor of electrochemically active phenazine derivatives, which plays an important role in anoxygenic conditions allowing the bacteria to produce energy for its growth. Phenazines also help to sustain the redox homeostasis as they act as electron acceptors to re-oxidize the accumulated nicotinamide adenine dinucleotide (NADH) (Jayapriya and Ramamurthy 2012). The chemical variations of the insulating interface across the cellular membrane can enhance the endogenous secretion of pyocyanin mediators in *Escherichia coli* (Hou et al. 2013) and *Pseudomonas aeruginosa* (Wang et al. 2013), thereby boosting the power output in the MFCs. Moreover, *Pseudomonas*-catalyzed MFCs support an exclusive prospect to syndicate the metabolic potential of microorganisms to convert oxidizable pollutants and with energy recovery.

8.4 REACTION MECHANISMS OF MICROBIAL FUEL CELLS

Cathode and anode are the two chambers in MFCs made up of glass, polycarbonate or Plexiglas. Microorganisms in the anodic chamber generate electrons and protons to metabolize the organic substrates, produce energy and support microbial growth (Das and Mangwani 2010). Protons and electrons travel towards the cathode through the PEM where they cause the reduction of oxygen to water. Oxygen serves as an electron recipient to complete the entire process. Oxygen is a sustainable, non-toxic compound and labeled as an ideal electron acceptor. A practical system is necessary to separate bacteria from oxygen because the latter is inhibitory for electricity production in the anode chamber. Hence, an anaerobic chamber is required for the anodic reaction. The anode is the chamber where bacteria grows and the cathode is the chamber where oxygen reacts with the electrons. A membrane separates biocatalyst from oxygen and allows only charges to be transferred between the anode and the cathode (Das and Mangwani 2010).

Two important factors that influence the functioning of MFCs are biological and electrochemical parameters. In the continuous systems, the rate of substrate loading is a biological parameter but power density and cell voltage are taken as chief electrochemical parameters. The performance of an MFC is also decided by various factors, which include: (i) supply of oxygen with its usage in the cathode chamber, (ii) oxidation of the substrate in the anode chamber, (iii) electron shuttle from anode section to anode surface and (v) penetrability of the PEM (Rahimnejad et al. 2015).

8.5 NOTABLE APPLICATIONS OF MICROBIAL FUEL CELLS

8.5.1 Electricity Generation and Wastewater Treatment

The major application of MFCs has been realized for bioelectricity generation subsequently treating the wastewater. A wide variety of microorganisms is involved in MFCs either as a single species or in consortia by the virtue of their unique metabolic potentials. Some of the chief substrates used by the microorganisms are sanitary wastes, wastewater generated out of food processing, poultry wastewater and corn stover (Rabaey et al. 2006). The growth promoters proliferate the development of bioelectrochemically active microorganisms during the wastewater treatment process. Wastewater treatment is more efficient when the sulfides are removed through microbial metabolism. The energy demands are highly curtailed on the treatment plant along with the reduction in the quantity of unfeasible sludge produced by the predominant anoxic atmosphere. The removal efficiency by MFCs is enhanced when they are connected in series to treat the leachate. The generation of electricity is the additional benefit associated with this process (Gálvez et al. 2009).

8.5.2 Biosensors

Organic matter can be monitored online when replaceable anaerobic consortia are used as biosensors in which the biological oxygen demand (BOD) in the wastewater is one of the chief parameters. Most of the methods are not feasible for on-line monitoring and regulation of biological wastewater treatment processes (Chang et al. 2005). The strength of the organic matter and the Coulombic yield of MFCs are linearly correlated, thus making the MFC a feasible BOD sensor. The BOD of a liquid stream can be better perceived by measuring the MFC's Coulombic yield. This is a feasible approach for a wide concentration range of organic matter in the wastewater (Kumlaghan et al. 2007).

8.5.3 Biofuel Production

Biohydrogen is generated in the MFCs as a biofuel used for the alternative of electricity (Nanda et al. 2017a; Nanda et al. 2017c). Biohydrogen production becomes feasible with minor changes in the MFCs. The MFCs are an alternative producer of biohydrogen when compared to the classical method of its production through photo-fermentation or dark fermentation (Sarangi and Nanda 2020). It is stated that by increasing the external potential at the cathode, microbial electrolytic cells can potentially generate methane and hydrogen. Thus, MFCs can generate biohydrogen and contribute towards fulfilling the energy demands in the future bioeconomy sector (Wagner et al. 2009).

8.6 FUTURE POSSIBILITIES IN MICROBIAL FUEL CELL TECHNOLOGY

Momentous improvements in the MFC technologies have been witnessed in recent decades. The practical utility and scale-up of MFC technologies encounter certain technical challenges of which the impediment of power generation is of major concern. The hindrance in the production of power is obstructed if the concentration of the substrate surpasses a specific level because there is a direct relationship between the substrate concentration and power generation by the MFC. The second issue that needs attention is the high internal resistance that consumes the generated power substantially, thereby curtailing the MFC output. As there is an involvement of EAB in the operation of MFCs, it is equally important to focus on improving the yield and transfer of electrons. Genetic modifications of EAB is an approach that can be further explored. Other research areas of attention are vital modifications in the electrode, optimization of process parameters and scale-up of the reactor.

The focus is also being bestowed on employing cost-effective and sustainable materials for constructing the MFCs. Achieving this target would help in scaling up of the process along with suitable disposal of toxic wastes being accumulated as electronic components, plastics and batteries. The performance of a MFC can be optimized if the suitable provisions are made for the removal of the organic load. All these strategies would boost up the power production by multifold levels. The possibility of employing cost-effective and non-toxic mediators should also be considered. It would be better if the need for using these artificial mediators can be subsided. The exploration of microbial species that can support direct conductance should also be encouraged.

8.7 CONCLUSIONS

Microbial fuel cells are an excellent approach to solve environmental issues and energy crises. A wide variety of microorganisms supports this process where they generate electrons from the oxidation of organic matter. The movement of these electrons generates electricity. Several mediators can be employed to facilitate this process. However, endogenous mediators of microbial origin can curtail the need for artificial mediators. Many efforts are being made to make this technology cost-effective and efficient. The last few years have witnessed the expansion of the scope of MFC utilization for bioelectricity production for specialized utilities. The MFC is a feasible approach for producing biofuels such as biohydrogen. Some lethal compounds can also be remediated with the help of MFCs technology.

REFERENCES

Bhatia, L., P.K. Sarangi, S. Nanda. 2020. Current advancements in microbial fuel cell technologies. In: *Biorefinery of Alternative Resources: Targeting Green Fuels and Platform Chemicals*; eds. S. Nanda, D.V.N. Vo, P.K. Sarangi; 477–494. Singapore: Springer Nature.

Borole, A.P., G. Reguera, B. Ringeisen, Z.W. Wang, Y. Feng, B.H. Kim. 2011. Electroactive biofilms: current status and future research needs. *Energy & Environmental Science* 4:4813–4834.

Chang, I.S., H. Moon, J.K. Jang, B.H. Kim. 2005. Improvement of a microbial fuel cell performance as a BOD sensor using respiratory inhibitors. *Biosensors and Bioelectronics* 20:1856–1859.

Das, S., N. Mangwani. 2010. Recent developments in microbial fuel cells: a review. *Journal of Scientific and Industrial Research* 69:727–731.

Gálvez, A., J. Greenman, I. Ieropoulos. 2009. Landfill leachate treatment with microbial fuel cells; scale-up through plurality. *Bioresource Technology* 100:5085–5091.

Gong, M., S. Nanda, H.N. Hunter, W. Zhu, A.K. Dalai, J.A. Kozinski. 2017a. Lewis acid catalyzed gasification of humic acid in supercritical water. *Catalysis Today* 291:13–23.

Gong, M., S. Nanda, M.J. Romero, W. Zhu, J.A. Kozinski. 2017b. Subcritical and supercritical water gasification of humic acid as a model compound of humic substances in sewage sludge. *The Journal of Supercritical Fluids* 119:130–138.

He, Z., S.D. Minteer, L.T. Angenent. 2005. Electricity generation from artificial wastewater using an upflow microbial fuel cell. *Environmental Science & Technology* 39:5262–5267.

Holmes, D.E., D.R. Bond, R.A. O'Neil, C.E. Reimers, L.R. Tender, D.R. Lovley. 2004. Microbial communities associated with electrodes harvesting electricity from a variety of aquatic sediments. *Microbial Ecology* 48:178–190.

Hou, H., X. Chen, A.W. Thomas. 2013. Conjugated oligo electrolytes increase power generation in E. coli microbial fuel cells. *Advanced Materials* 25:1593–1597.

Ieropoulos, I.A., P. Ledezma, A. Stinchcombe, G. Papaharalabos, C. Melhuish, J. Greenman. 2013. Waste to real energy: the first MFC powered mobile phone. *Physical Chemistry Chemical Physics* 15:15312–15316.

Jayapriya, J., V. Ramamurthy. 2012. Use of non-native phenazines to improve the performance of *Pseudomonas aeruginosa* MTCC 2474 catalysed fuel cells. *Bioresource Technology* 124:23–28.

Kumlaghan, A., J. Liu, P. Thavarungkul, P. Kanatharana, B. Mattiasson. 2007. Microbial fuel cell-based biosensor for fast analysis of biodegradable organic matter. *Biosensors and Bioelectronics* 22:2939–2944.

Larrosa-Guerrero, A., K. Scott, I.M. Head, F. Mateo, A. Ginesta, C. Godinez. 2010. Effect of temperature on the performance of microbial fuel cells. *Fuel* 89:3985–3994.

Nanda, S., A.K. Dalai, I. Gökalp, J.A. Kozinski. 2016a. Valorization of horse manure through catalytic supercritical water gasification. *Waste Management* 52:147–158.

Nanda, S., A.K. Dalai, J.A. Kozinski. 2016b. Supercritical water gasification of timothy grass as an energy crop in the presence of alkali carbonate and hydroxide catalysts. *Biomass & Bioenergy* 95:378–387.

Nanda, S., J. Isen, A.K. Dalai, J.A. Kozinski. 2016c. Gasification of fruit wastes and agro-food residues in supercritical water. *Energy Conversion and Management* 110:296–306.

Nanda, S., J. Mohammad, S.N. Reddy, J.A. Kozinski, A.K. Dalai. 2014. Pathways of lignocellulosic biomass conversion to renewable fuels. *Biomass Conversion and Biorefinery* 4:157–191.

Nanda, S., J.A. Kozinski, A.K. Dalai. 2016d. Lignocellulosic biomass: a review of conversion technologies and fuel products. *Current Biochemical Engineering* 3:24–36.

Nanda, S., K. Li, N. Abatzoglou, A.K. Dalai, J.A. Kozinski. 2017a. Advancements and confinements in hydrogen production technologies. In: *Bioenergy Systems for the Future*; eds. F. Dalena, A. Basile, C. Rossi; 373–418. Cambridge: Woodhead Publishing, Elsevier.

Nanda, S., M. Gong, H.N. Hunter, A.K. Dalai, I. Gökalp, J.A. Kozinski. 2017b. An assessment of pinecone gasification in subcritical, near-critical and supercritical water. *Fuel Processing Technology* 168:84–96.

Nanda, S., R. Azargohar, A.K. Dalai, J.A. Kozinski. 2015a. An assessment on the sustainability of lignocellulosic biomass for biorefining. *Renewable and Sustainable Energy Reviews* 50:925–941.

Nanda, S., R. Rana, Y. Zheng, J.A. Kozinski, A.K. Dalai. 2017c. Insights on pathways for hydrogen generation from ethanol. *Sustainable Energy & Fuels* 1:1232–1245.

Nanda, S., S.N. Reddy, H.N. Hunter, I.S. Butler, J.A. Kozinski. 2015b. Supercritical water gasification of lactose as a model compound for valorization of dairy industry effluents. *Industrial & Engineering Chemistry Research* 54:9296–9306.

Nanda, S., S.N. Reddy, S.K. Mitra, J.A. Kozinski. 2016e. The progressive routes for carbon capture and sequestration. *Energy Science & Engineering* 4:99–122.

Okolie, J.A., R. Rana, S. Nanda, A.K. Dalai, J.A. Kozinski. 2019. Supercritical water gasification of biomass: a state-of-the-art review of process parameters, reaction mechanisms and catalysis. *Sustainable Energy & Fuels* 3:578–598.

Okolie, J.A., S. Nanda, A.K. Dalai, F. Berruti, J.A. Kozinski. 2020a. A review on subcritical and supercritical water gasification of biogenic, polymeric and petroleum wastes to hydrogen-rich synthesis gas. *Renewable and Sustainable Energy Reviews* 119:109546.

Okolie, J.A., S. Nanda, A.K. Dalai, J.A. Kozinski. 2020b. Chemistry and specialty industrial applications of lignocellulosic biomass. *Waste and Biomass Valorization*. Doi: 10.1007/s12649-020-01123-0.

Osman, M.H., A.A. Shah, F.C. Walsh. 2010. Recent progress and continuing challenges in biofuel cells. Part II: Microbial. *Biosensors and Bioelectronics* 26:953–963.

Pant, D., G.V. Bogaert, L. Diels, K. Vanbroekhoven. 2010. A review of the substrates used in microbial fuel cells (MFCs) for sustainable energy production. *Bioresource Technology* 101:1533–1543.

Parakh, P.D., S. Nanda, J.A. Kozinski. 2020. Eco-friendly transformation of waste biomass to biofuels. *Current Biochemical Engineering* 6:120–134.

Rabaey, K., K. Van De Sompel, L. Maignien, N. Boon, P. Aelterman, P. Clauwaert, L. De Schamphelaire, H.T. Pham, J. Vermeulen, M. Verhaege, P. Lens, W. Verstraete. 2006. Microbial fuel cells for sulfide removal. *Environmental Science & Technology* 40:5218–5224.

Rahimnejad, M., A. Adhami, S. Darvari, A. Zirepour, S.E. Oh. 2015. Microbial fuel cell as new technology for bioelectricity generation: a review. *Alexandria Engineering Journal* 54:745–756.

Rana, R., S. Nanda, A. Maclennan, Y. Hu, J.A. Kozinski, A.K. Dalai. 2019. Comparative evaluation for catalytic gasification of petroleum coke and asphaltene in subcritical and supercritical water. *Journal of Energy Chemistry* 31:107–118.

Rana, R., S. Nanda, J.A. Kozinski, A.K. Dalai. 2018. Investigating the applicability of Athabasca bitumen as a feedstock for hydrogen production through catalytic supercritical water gasification. *Journal of Environmental Chemical Engineering* 6:182–189.

Rana, R., S. Nanda, S.N. Reddy, A.K. Dalai, J.A. Kozinski, I. Gökalp. 2020. Catalytic gasification of light and heavy gas oils in supercritical water. *Journal of the Energy Institute* 93, 2025–2032.

Reddy, S.N., S. Nanda, J.A. Kozinski. 2016. Supercritical water gasification of glycerol and methanol mixtures as model waste residues from biodiesel refinery. *Chemical Engineering Research and Design* 113:17–27.

Sarangi, P.K., S. Nanda. 2018. Recent developments and challenges of acetone-butanol-ethanol fermentation. In: *Recent Advancements in Biofuels and Bioenergy Utilization*; eds. P.K. Sarangi, S. Nanda, P. Mohanty; 111–123. Singapore: Springer Nature.

Sarangi, P.K., S. Nanda. 2019. Recent advances in consolidated bioprocessing for microbe-assisted biofuel production. In: *Fuel Processing and Energy Utilization*; eds. S. Nanda, P.K. Sarangi, D.V.N. Vo; 141–157. Florida, USA: CRC Press.

Sarangi, P.K., S. Nanda. 2020. Biohydrogen production through dark fermentation. *Chemical Engineering & Technology* 43:601–612.

Sarangi, P.K., S. Nanda, D.V.N. Vo. 2020. Technological advancements in the production and application of biomethanol. In: *Biorefinery of Alternative Resources: Targeting Green Fuels and Platform Chemicals*; eds. S. Nanda, D.V.N. Vo, P.K. Sarangi; 127–139. Singapore: Springer Nature.

Shen, H.B., X.Y. Yong, Y.L. Chen, Z.H. Liao, R.W. Si, J. Zhou, S.Y. Wang, Y.C. Yong, P.K. Ouyang, T. Zheng. 2014. Enhanced bioelectricity generation by improving pyocyanin production and membrane permeability through sophorolipid addition in *Pseudomonas aeruginosa* inoculated microbial fuel cells. *Bioresource Technology* 167:490–494.

Singh, A., S. Nanda, F. Berruti. 2020. A review of thermochemical and biochemical conversion of *Miscanthus* to biofuels. In: *Biorefinery of Alternative Resources: Targeting Green Fuels and Platform Chemicals*; eds. S. Nanda, D.V.N. Vo, P.K. Sarangi; 195–220. Singapore: Springer Nature.

Tremouli, A., M. Martinos, G. Lyberatos. 2016. The effects of salinity, pH and temperature on the performance of a microbial fuel cell. *Waste and Biomass Valorization* 8:2037–2043.

Wagner, R.C., J.M. Regan, S.E. Oh, Y. Zuo, B.E. Logan. 2009. Hydrogen and methane production from swine wastewater using microbial electrolysis cells. *Water Research* 43:1480–1488.

Wang, V.B., S. Chua, B. Cao. 2013. Engineering PQS biosynthesis pathway for enhancement of bioelectricity producing *Pseudomonas aeruginosa* microbial fuel cells. *PLOS One* 8:5.

Wang, V.B., S.L. Chua, Z. Cai. 2014. A stable synergistic microbial consortium for simultaneous azo dye removal and bioelectricity generation. *Bioresource Technology* 155:71–76.

Xu, B., Z. Ge, Z. He. 2015. Sediment microbial fuel cells for wastewater treatment: challenges and opportunities. *Environmental Science: Water Research and Technology* 1:279–284.

Index